Nonlinear Dynamics: A Primer

This book provides a systematic and comprehensive introduction to the study of nonlinear dynamical systems, in both discrete and continuous time, for nonmathematical students and researchers working in applied fields including economics, physics, engineering, biology, statistics and linguistics. It includes a review of linear systems and an introduction to the classical theory of stability, as well as chapters on the stability of invariant sets, bifurcation theory, chaotic dynamics and the transition to chaos. In the final chapters the authors approach the study of dynamical systems from a measure-theoretical point of view, comparing the main notions and results to their counterparts in the geometrical or topological approach. Finally, they discuss the relations between deterministic systems and stochastic processes.

The book is part of a complete teaching unit. It includes a large number of pencil and paper exercises, and an associated website offers, free of charge, a Windows-compatible software program, a workbook of computer exercises coordinated with chapters and exercises in the book, answers to selected book exercises, and further teaching material.

ALFREDO MEDIO is Professor of Mathematical Economics at the University 'Ca' Foscari' of Venice and Director of the International Centre of Economics and Finance at the Venice International University. He has published widely on applications of dynamical system theory; his books include *Chaotic Dynamics: Theory and Applications to Economics* (Cambridge University Press, 1992).

MARJI LINES is Associate Professor of Economics at the University of Udine, Italy where she has taught courses in microeconomics, mathematical economics and econometrics. She has published articles on economic theory and environmental economics in journals and collections.

Nonlinear Dynamics

A Primer

ALFREDO MEDIO

MARJI LINES

CAMBRIDGE
UNIVERSITY PRESS

CAMBRIDGE
UNIVERSITY PRESS

University Printing House, Cambridge CB2 8BS, United Kingdom

Cambridge University Press is part of the University of Cambridge.

It furthers the University's mission by disseminating knowledge in the pursuit of education, learning and research at the highest international levels of excellence.

www.cambridge.org
Information on this title: www.cambridge.org/9780521551861

© Alfredo Medio and Marji Lines 2001

First published 2001

A catalogue record for this publication is available from the British Library

ISBN 978-0-521-55186-1 Hardback
ISBN 978-0-521-55874-7 Paperback

To the memory of my father
who taught me to love books

To my mother and father

Contents

Preface

Over the years we have had the rare opportunity to teach small classes of intelligent and strongly motivated economics students who found nonlinear dynamics inspiring and wanted to know more. This book began as an attempt to organise our own ideas on the subject and give the students a fairly comprehensive but reasonably short introduction to the relevant theory. Cambridge University Press thought that the results of our efforts might have a more general audience.

The theory of nonlinear dynamical systems is technically difficult and includes complementary ideas and methods from many different fields of mathematics. Moreover, as is often the case for a relatively new and fast growing area of research, coordination between the different parts of the theory is still incomplete, in spite of several excellent monographs on the subject. Certain books focus on the geometrical or topological aspects of dynamical systems, others emphasise their ergodic or probabilistic properties. Even a cursory perusal of some of these books will show very significant differences not only in the choice of content, but also in the characterisations of some fundamental concepts. (This is notoriously the case for the concept of attractor.)

For all these reasons, any introduction to this beautiful and intellectually challenging subject encounters substantial difficulties, especially for non-mathematicians, as are the authors and the intended readers of this book. We shall be satisfied if the book were to serve as an access to the basic concepts of nonlinear dynamics and thereby stimulate interest on the part of students and researchers, in the physical as well as the social sciences, with a basic mathematical background and a good deal of intellectual curiosity.

The book includes those results in dynamical system theory that we deemed most relevant for applications, often accompanied by a common-sense interpretation. We have also tried, when necessary, to eliminate the

confusion arising from the lack of consistent and universally accepted definitions of some concepts. Full mathematical proofs are usually omitted and
the reader is referred either to the original sources or to some more recent
version of the proofs (with the exception of some 'canonical' theorems whose
discussion can be found virtually in any textbook on the subject).

We devote an unusually large space to the discussion of stability, a subject
that in the past played a central role in the theory of differential equations
and related applied research. The fundamental monographs on stability
were published in the 1960s or early 1970s yet there is surprisingly little
reference to them in modern contemporary research on dynamical systems.
We have tried to establish a connection between the classical theory of
stability and the more recent discussions of attracting sets and attractors.

Although the word 'chaos' does not appear in the title, we have dedicated
substantial attention to chaotic sets and attractors as well as to 'routes to
chaos'. Moreover, the geometric or topological properties of chaotic dynamics are compared to their measure-theoretic counterparts.

We provide precise definitions of some basic notions such as neighbourhood, boundary, closure, interior, dense set and so on, which mathematicians might find superfluous but, we hope, will be appreciated by students
from other fields.

At an early stage in the preparation of this book, we came to the conclu-
sion that, within the page limit agreed upon with the publisher, we could not
cover both theory and applications. We squarely opted for the former. The
few applications discussed in detail belong to economics where our compara-
tive advantages lie, but we emphasised the multi-purpose techniques rather
than the specificities of the selected models.

The book includes about one hundred exercises, most of them easy and
requiring only a short time to solve.

In 1992, Cambridge University Press published a book on Chaotic Dynamics by the first author, which contained the basic concepts of chaos theory necessary to perform and understand numerical simulations of difference
and differential equations. That book included a user-friendly software program called DMC (Dynamical Models Cruncher). A refurbished, enlarged
and Windows-compatible version of the program is available, at no cost,
from the webpage

<http://uk.cambridge.org/economics/catalogue/0521558743>

along with a workbook of computer exercises coordinated with the 'paper
and pencil' exercises found in the book. The webpage will also be used to
circulate extra exercises, selected solutions and, we hope, comments and
criticisms by readers.

We take this opportunity to give our warm thanks to those who, in different capacities and at different times, helped us complete this book.

Laura Gardini, Hans-Walter Lorenz, Ami Radunskaya, Marcellino Gaudenzi, Gian Italo Bischi, Andrea Sgarro, Sergio Invernizzi and Gabriella Caristi, commented on preliminary versions of parts of the book or gave their advice on specific questions and difficulties. We did not always follow their suggestions, and, at any rate, the responsibility for all remaining errors and misunderstandings remains entirely with us. Thanks also to Eric Kostelich, Giancarlo Benettin and Luigi Galgani who, in various conversations, helped us clarify specific issues.

At Cambridge University Press we would like to thank Patrick McCartan, for suggesting the idea; Ashwin Rattan, Economics Editor, for his support and patience; Alison Woollatt for her TeX advice. Thanks also to Barbara Docherty for her excellent editing.

The authors also gratefully acknowledge financial help from the Italian Ministry of the University (MURST) and the Italian National Council of Research (CNR).

Alfredo Medio and Marji Lines
Venice, November 2000

1

Statics and dynamics: some elementary concepts

Dynamics is the study of the movement through time of variables such as heartbeat, temperature, species population, voltage, production, employment, prices and so forth.

This is often achieved by means of equations linking the values of variables at different, uniformly spaced instants of time, i.e., **difference equations**, or by systems relating the values of variables to their time derivatives, i.e., **ordinary differential equations**. Dynamical phenomena can also be investigated by other types of mathematical representations, such as partial differential equations, lattice maps or cellular automata. In this book, however, we shall concentrate on the study of systems of difference and differential equations and their dynamical behaviour.

In the following chapters we shall occasionally use models drawn from economics to illustrate the main concepts and methods. However, in general, the mathematical properties of equations will be discussed independently of their applications.

1.1 A static problem

To provide a first, broad idea of the problems posed by dynamic *vis-à-vis* static analysis, we shall now introduce an elementary model that could be labelled as 'supply-demand-price interaction in a single market'. Our model considers the quantities supplied and demanded of a single good, defined as functions of a single variable, its price, p. In economic parlance, this would be called partial analysis since the effect of prices and quantities determined in the markets of all other goods is neglected. It is assumed that the demand function $D(p)$ is decreasing in p (the lower the price, the greater the amount that people wish to buy), while the supply function $S(p)$ is increasing in p (the higher the price, the greater the amount that people wish to supply).

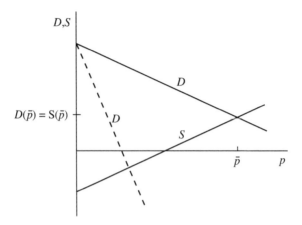

Fig. 1.1 The static partial equilibrium model

For example, in the simpler, linear case, we have:

$$D(p) = a - bp$$
$$S(p) = -m + sp$$
(1.1)

and a, b, m and s are positive constants. Only nonnegative values of these variables are economically meaningful, thus we only consider $D, S, p \geq 0$. The **economic equilibrium condition** requires that the market of the good clears, that is demand equals supply, namely:

$$D(p) = S(p)$$
(1.2)

or

$$a - bp = -m + sp.$$

STATIC SOLUTION Mathematically, the solution to our problem is the value of the variable that solves (1.2) (in this particular case, a linear equation). Solving (1.2) for p we find:

$$\bar{p} = \frac{a + m}{b + s}$$

where $\bar{p}$ is usually called the **equilibrium price** (see figure 1.1).[1] We call the problem *static* since no reference is made to time or, if you prefer,

[1] The demand curve D' in figure 1.1 is provided to make the point that, with no further constraints on parameter values, the equilibrium price could imply negative equilibrium quantities of supply and demand. To eliminate this possibility we further assume that $0 < m/s \leq a/b$, as is the case for the demand curve D.

everything happens at the same time. Notice that, even though the static model allows us to find the equilibrium price of the good, it tells us nothing about what happens if the actual price is different from its equilibrium value.

1.2 A discrete-time dynamic problem

The introduction of dynamics into the model requires that we replace the equilibrium condition (1.2) with some hypothesis concerning the behaviour of the system off-equilibrium, i.e., when demand and supply are not equal. For this purpose, we assume the most obvious mechanism of price adjustment: over a certain interval of time, the price increases or decreases in proportion to the excess of demand over supply, $(D - S)$ (for short, **excess demand**). Of course, excess demand can be a positive or a negative quantity. Unless the adjustment is assumed to be instantaneous, prices must now be dated and p_n denotes the price of the good at time n, time being measured at equal intervals of length h. Formally, we have

$$p_{n+h} = p_n + h\theta[D(p_n) - S(p_n)]. \tag{1.3}$$

Since h is the period of time over which the adjustment takes place, θ can be taken as a measure of the speed of price response to excess demand. For simplicity, let us choose $h = 1$, $\theta = 1$. Then we have, making use of the demand and supply functions (1.1),

$$p_{n+1} = a + m + (1 - b - s)p_n. \tag{1.4}$$

In general, a solution of (1.4) is *a function of time $p(n)$* (with n taking discrete, integer values) that satisfies (1.4).[2]

DYNAMIC SOLUTION To obtain the full dynamic solution of (1.4), we begin by setting $\alpha = a + m$, $\beta = (1 - b - s)$ to obtain

$$p_{n+1} = \alpha + \beta p_n. \tag{1.5}$$

To solve (1.5), we first set it in a canonical form, with all time-referenced terms of the variable on the left hand side (LHS), and all constants on the right hand side (RHS), thus:

$$p_{n+1} - \beta p_n = \alpha. \tag{1.6}$$

Then we proceed in steps as follows.

---

[2] We use the forms p_n and $p(n)$ interchangeably, choosing the latter whenever we prefer to emphasise that p is a function of n.

STEP 1 We solve the **homogeneous equation**, which is formed by setting the RHS of (1.6) equal to 0, namely:

$$p_{n+1} - \beta p_n = 0. \tag{1.7}$$

It is easy to see that a function of time $p(n)$ satisfying (1.7) is $p(n) = C\beta^n$, with C an arbitrary constant. Indeed, substituting in (1.7), we have

$$C\beta^{n+1} - \beta C\beta^n = C\beta^{n+1} - C\beta^{n+1} = 0.$$

STEP 2 We find a **particular solution** of (1.6), assuming that it has a form similar to the RHS in the general form. Since the latter is a constant, set $p(n) = k$, k a constant, and substitute it into (1.6), obtaining

$$k - \beta k = \alpha$$

so that

$$k = \frac{\alpha}{1 - \beta} = \frac{a + m}{b + s} = \bar{p} \qquad \text{again!}$$

It follows that the $p(n) = \bar{p}$ is a solution to (1.6) and the constant (or stationary) solution of the dynamic problem is simply the solution of the static problem of section 1.1.

STEP 3 Since (1.6) is linear, the sum of the homogeneous and the particular solution is again a solution,[3] called the **general solution**. This can be written as

$$p(n) = \bar{p} + C\beta^n. \tag{1.8}$$

The arbitrary constant C can now be expressed in terms of the initial condition. Putting $p(0) \equiv p_0$, and solving (1.8) for C we have

$$p_0 = \bar{p} + C\beta^0 = \bar{p} + C$$

whence $C = p_0 - \bar{p}$, that is, the difference between the initial and equilibrium values of p. The general solution can now be re-written as

$$p(n) = \bar{p} + (p_0 - \bar{p})\beta^n. \tag{1.9}$$

Letting n take integer values $1, 2, \ldots$, from (1.9) we can generate a sequence of values of p, a 'history' of that variable (and consequently, a history of quantities demanded and supplied at the various prices), once its value at any arbitrary instant of time is given. Notice that, since the function $p_{n+1} =$

---

[3] This is called the *superposition principle* and is discussed in detail in chapter 2 section 2.1.

$f(p_n)$ is **invertible**, i.e., the function f^{-1} is well defined, $p_{n-1} = f^{-1}(p_n)$ also describes the past history of p.

The value of p at each instant of time is equal to the sum of the equilibrium value (the solution to the static problem which is also the particular, stationary solution) and the initial disequilibrium $(p_0 - \bar{p})$, amplified or dampened by a factor β^n. There are therefore two basic cases:

(i) $|\beta| > 1$. Any nonzero deviation from equilibrium is amplified in time, the equilibrium solution is unstable and as $n \to +\infty$, p_n asymptotically tends to $+\infty$ or $-\infty$.

(ii) $|\beta| < 1$. Any nonzero deviation is asymptotically reduced to zero, $p_n \to \bar{p}$ as $n \to +\infty$ and the equilibrium solution is consequently stable.

First-order, discrete-time equations (where the order is determined as the difference between the extreme time indices) can also have fluctuating behaviour, called **improper oscillations**,[4] owing to the fact that if $\beta < 0$, β^n will be positive or negative according to whether n is even or odd. Thus the sign of the adjusting component of the solution, the second term of the RHS of (1.9), oscillates accordingly. Improper oscillations are dampened if $\beta > -1$ and explosive if $\beta < -1$.

In figure 1.2 we have two representations of the motion of p through time. In figure 1.2(a) we have a line defined by the solution equation (1.5), and the bisector passing through the origin which satisfies the equation $p_{n+1} = p_n$. The intersection of the two lines corresponds to the constant, equilibrium solution. To describe the off-equilibrium dynamics of p, we start on the abscissa from an initial value $p_0 \neq \bar{p}$. To find p_1, we move vertically to the solution line and sidewise horizontally to the ordinate. To find p_2, we first reflect the value of p_1 by moving horizontally to the bisector and then vertically to the abscissa. From the point p_1, we repeat the procedure proposed for p_0 (up to the solution line, left to the ordinate), and so on and so forth. The procedure can be simplified by omitting the intermediate step and simply moving up to the solution line and sidewise to the bisector, up again, and so on, as indicated in figure 1.2(a). It is obvious that for $|\beta| < 1$, at each iteration of the procedure the initial deviation from equilibrium is diminished again, see figure 1.2(b). For example, if $\beta = 0.7$, we have $\beta^2 = 0.49$, $\beta^3 = 0.34, \ldots, \beta^{10} \approx 0.03, \ldots$) and the equilibrium solution is approached asymptotically.

The reader will notice that stability of the system and the possibility

---

[4]The term *improper* refers to the fact that in this case oscillations of variables have a 'kinky' form that does not properly describe the smoother ups and downs of real variables. We discuss *proper* oscillations in chapter 3.

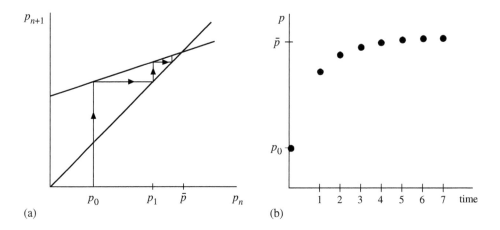

Fig. 1.2 Convergence to $\bar{p}$ in the discrete-time partial equilibrium model

of oscillatory behaviour depends entirely on β and therefore on the two parameters b and s, these latter denoting respectively the slopes of the demand and supply curves. The other two parameters of the system, a and m, determine α and consequently they affect only the equilibrium value $\bar{p}$. We can therefore completely describe the dynamic characteristics of the solution (1.9) over the parameter space (b, s). The boundary between stable and unstable behaviour is given by $|\beta| = 1$, and convergence to equilibrium is guaranteed for

$$-1 < \beta < 1$$
$$2 > b + s > 0.$$

The assumptions on the demand and supply functions imply that $b, s > 0$. Therefore, the stability condition is $(b + s) < 2$, the stability boundary is the line $(b + s) = 2$, as represented in figure 1.3. Next, we define the curve $\beta = 1 - (b + s) = 0$, separating the zone of monotonic behaviour from that of improper oscillations, which is also represented in figure 1.3. Three zones are labelled according to the different types of dynamic behaviour, namely: convergent and monotonic; convergent and oscillatory; divergent and oscillatory. Since $b, s > 0$, we never have the case $\beta > 1$, corresponding to divergent, nonoscillatory behaviour.

If $|\beta| > 1$ any initial difference $(p_0 - \bar{p})$ is amplified at each step. In this model, we can have $|\beta| > 1$ if and only if $\beta < -1$. Instability, then, is due to **overshooting**. Any time the actual price is, say, too low and there is positive excess demand, the adjustment mechanism generates a change in the price in the 'right' direction (the price rises) but the change is too large.

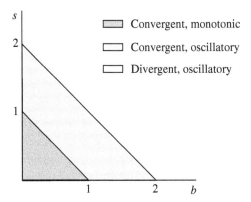

Fig. 1.3 Parameter space for the discrete-time partial equilibrium model

After the correction, the new price is too high (negative excess demand) and the discrepancy from equilibrium is larger than before. A second adjustment follows, leading to another price that is far too low, and so on. We leave further study of this case to the exercises at the end of this chapter.

1.3 A continuous-time dynamic problem

We now discuss our simple dynamical model in a continuous-time setting. Let us consider, again, the price adjustment equation (1.3) (with $\theta = 1$, $h > 0$) and let us adopt the notation $p(n)$ so that

$$p(n + h) = p(n) + h\left(D[p(n)] - S[p(n)]\right).$$

Dividing this equation throughout by h, we obtain

$$\frac{p(n + h) - p(n)}{h} = D[p(n)] - S[p(n)]$$

whence, taking the limit of the LHS as $h \to 0$, and recalling the definition of a derivative, we can write

$$\frac{dp(n)}{dn} = D[p(n)] - S[p(n)].$$

Taking the interval h to zero is tantamount to postulating that time is a continuous variable. To signal that time is being modelled differently we substitute the time variable $n \in \mathbb{Z}$ with $t \in \mathbb{R}$ and denote the value of p at time t simply by p, using the extended form $p(t)$ when we want to emphasise that price is a function of time. We also make use of the efficient Newtonian notation $dx(t)/dt = \dot{x}$ to write the price adjustment mechanism as

$$\frac{dp}{dt} = \dot{p} = D(p) - S(p) = (a + m) - (b + s)p. \tag{1.10}$$

Equation (1.10) is an ordinary differential equation relating the values of the variable p at a given time t to its first derivative with respect to time at the same moment. It is **ordinary** because the solution $p(t)$ is a function of a single independent variable, time. Partial differential equations, whose solutions are functions of more than one independent variable, will not be treated in this book, and when we refer to differential equations we mean ordinary differential equations.

DYNAMIC SOLUTION The dynamic problem is once again that of finding a function of time $p(t)$ such that (1.10) is satisfied for an arbitrary initial condition $p(0) \equiv p_0$.

As in the discrete-time case, we begin by setting the equation in canonical form, with all terms involving the variable or its time derivatives on the LHS, and all constants or functions of time (if they exist) on the RHS, thus

$$\dot{p} + (b + s)p = a + m. \tag{1.11}$$

Then we proceed in steps as follows.

STEP 1 We solve the homogeneous equation, formed by setting the RHS of (1.11) equal to 0, and obtain

$$\dot{p} + (b + s)p = 0 \text{ or } \dot{p} = -(b + s)p. \tag{1.12}$$

If we now integrate (1.12) by separating variables, we have

$$\int \frac{dp}{p} = -(b + s) \int dt$$

whence

$$\ln p(t) = -(b + s)t + A$$

where A is an arbitrary integration constant. Taking now the antilogarithm of both sides and setting $e^A = C$, we obtain

$$p(t) = Ce^{-(b+s)t}.$$

STEP 2 We look for a particular solution to the nonhomogeneous equation (1.11). The RHS is a constant so we try $p = k$, k a constant and consequently $\dot{p} = 0$. Therefore, we have

$$\dot{p} = 0 = (a + m) - (b + s)k$$

whence

$$k = \frac{a + m}{b + s} = \bar{p}.$$

Once again the solution to the static problem turns out to be a special (stationary) solution to the corresponding dynamic problem.

STEP 3 Since (1.12) is linear, the general solution can be found by summing the particular solution and the solution to the homogeneous equation, thus

$$p(t) = \bar{p} + Ce^{-(b+s)t}.$$

Solving for C in terms of the initial condition, we find

$$p(0) \equiv p_0 = \bar{p} + C \text{ and } C = (p_0 - \bar{p}).$$

Finally, the complete solution to (1.10) in terms of time, parameters, initial and equilibrium values is

$$p(t) = \bar{p} + (p_0 - \bar{p})e^{-(b+s)t}. \tag{1.13}$$

As in the discrete case, the solution (1.13) can be interpreted as the sum of the equilibrium value and the initial deviation of the price variable from equilibrium, amplified or dampened by the term $e^{-(b+s)t}$. Notice that in the continuous-time case, a solution to a differential equation $\dot{p} = f(p)$ always determines both the future and the past history of the variable p, independently of whether the function f is invertible or not. In general, we can have two main cases, namely:

(i) $(b + s) > 0$ Deviations from equilibrium tend asymptotically to zero as $t \rightarrow +\infty$.

(ii) $(b + s) < 0$ Deviations become indefinitely large as $t \rightarrow +\infty$ (or, equivalently, deviations tend to zero as $t \rightarrow -\infty$).

Given the assumptions on the demand and supply functions, and therefore on b and s, the explosive case is excluded for this model. If the initial price is below its equilibrium value, the adjustment process ensures that the price increases towards it, if the initial price is above equilibrium, the price declines to it. (There can be no overshooting in the continuous-time case.) In a manner analogous to the procedure for difference equations, the equilibria

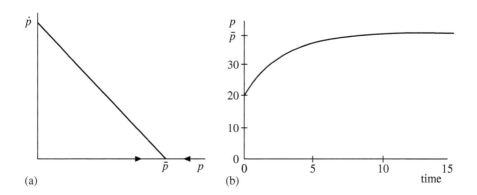

Fig. 1.4 The continuous-time partial equilibrium model

of differential equations can be determined graphically in the plane $(p, \dot{p})$ as suggested in figure 1.4(a). Equilibria are found at points of intersection of the line defined by (1.10) and the abscissa, where $\dot{p} = 0$. Convergence to equilibrium from an initial value different from the equilibrium value is shown in figure 1.4(b).

Is convergence likely for more general economic models of price adjustment, where other goods and income as well as substitution effects are taken into consideration? A comprehensive discussion of these and other related microeconomic issues is out of the question in this book. However, in the appendixes to chapter 3, which are devoted to a more systematic study of stability in economic models, we shall take up again the question of convergence to or divergence from economic equilibrium.

We would like to emphasise once again the difference between the discrete-time and the continuous-time formulation of a seemingly identical problem, represented by the two equations

$$p_{n+1} - p_n = (a + m) - (b + s)p_n \qquad (1.4)$$
$$\dot{p} = (a + m) - (b + s)p. \qquad (1.10)$$

Whereas in the latter case $(b+s) > 0$ is a sufficient (and necessary) condition for convergence to equilibrium, stability of (1.4) requires that $0 < (b+s) < 2$, a tighter condition.

This simple fact should make the reader aware that a naive translation of a model from discrete to continuous time or vice versa may have unsuspected consequences for the dynamical behaviour of the solutions.

1.4 Flows and maps

To move from the elementary ideas and examples considered so far to a more general and systematic treatment of the matter, we need an appropriate mathematical framework, which we introduce in this section. When necessary, the most important ideas and methods will be discussed in greater detail in the following chapters. For the sake of presentation, we shall begin with continuous-time systems of differential equations, which typically take the canonical form

$$\frac{dx}{dt} = \dot{x} = f(x) \tag{1.14}$$

where f is a function with domain U, an open subset of $\mathbb{R}^m$, and range $\mathbb{R}^m$ (denoted by $f: U \to \mathbb{R}^m$). The vector[5] $x = (x_1, x_2, \ldots, x_m)^T$ denotes the physical variables to be studied, or some appropriate transformations of them; $t \in \mathbb{R}$ indicates time. The variables x_i are sometimes called 'dependent variables' whereas t is called the 'independent variable'.

Equation (1.14) is called **autonomous** when the function f does not depend on time directly, but only through the state variable x. In this book we shall be mainly concerned with this type of equation, but in our discussions of stability in chapters 3 and 4 we shall have something to say about nonautonomous equations as well.

The space $\mathbb{R}^m$, or an appropriate subspace of dependent variables — that is, variables whose values specify the state of the system — is referred to as the **state space**. It is also known as the **phase space** or, sometimes, the **configuration space**, but we will use only the first term. Although for most of the problems encountered in this book the state space is the Euclidean space, we occasionally discuss dynamical systems different from $\mathbb{R}^m$, such as the unit circle. The circle is a one-dimensional object embedded in a two-dimensional Euclidean space. It is perhaps the simplest example of a kind of set called **manifold**. Roughly speaking, a manifold is a set which locally looks like a piece of $\mathbb{R}^m$. A more precise definition is deferred to appendix C of chapter 3, p. 98.

In simple, low-dimensional graphical representations of the state space the direction of motion through time is usually indicated by arrows pointing to the future. The enlarged space in which the time variable is explicitly

---

[5] Recall that the **transposition operator**, or transpose, designated by ^{T}, when applied to a row vector, returns a column vector and vice versa. When applied to a matrix, the operator interchanges rows and columns. Unless otherwise indicated, vectors are column vectors.

considered is called the **space of motions**. Schematically, we have

$$\mathbb{R} \quad \times \quad \mathbb{R}^m \quad = \quad \mathbb{R}^{1+m}$$
$$\downarrow \qquad\qquad \downarrow \qquad\qquad \downarrow$$
$$\text{time} \qquad \text{state} \qquad \text{space of}$$
$$\text{space} \qquad \text{motions}$$

The function f defining the differential equation (1.14) is also called a **vector field**, because it assigns to each point $x \in U$ a velocity vector $f(x)$. A solution of (1.14) is often written as a function $x(t)$, where $x : I \to \mathbb{R}^m$ and I is an interval of $\mathbb{R}$. If we want to specifically emphasise the solution that, at the initial time t_0, passes through the initial point x_0, we can write $x(t; t_0, x_0)$, where $x(t_0; t_0, x_0) = x(t_0) = x_0$. We follow the practice of setting $t_0 = 0$ when dealing with autonomous systems whose dynamical properties do not depend on the choice of initial time.

remark 1.1 In applications, we sometimes encounter differential equations of the form

$$\frac{d^m x}{dt^m} = F\left(x, \frac{dx}{dt}, \dots, \frac{d^{m-1} x}{dt^{m-1}}\right) \qquad x \in \mathbb{R} \tag{1.15}$$

where $d^k x/dt^k$ denotes the kth derivative of x with respect to time. Equation (1.15) is an autonomous, ordinary differential equation of order m, where m is the highest order of differentiation with respect to time appearing in the equation. It can always be put into the canonical form (1.14) by introducing an appropriate number of auxiliary variables. Specifically, put

$$\frac{d^k x}{dt^k} = z_{k+1}, \qquad 0 \le k \le m - 1$$

(where, for $k = 0$, $d^k x/dt^k = x$) so that

$$\dot{z}_k = z_{k+1}, \qquad 1 \le k \le m - 1$$
$$\dot{z}_m = F(z_1, \dots, z_m).$$

If we now denote by $z \in \mathbb{R}^m$ the m-dimensional vector $(z_1, \dots, z_m)^T$ we can write:

$$\dot{z} = f(z)$$

where $f(z) = [z_2, \dots, z_m, F(z_1, \dots, z_m)]^T$.

We can also think of solutions of differential equations in a different manner which is now prevalent in dynamical system theory and will be very helpful for understanding some of the concepts discussed in the following chapters.

If we denote by $\phi_t(x) = \phi(t, x)$ the state in $\mathbb{R}^m$ reached by the system at time t starting from x, then the totality of solutions of (1.14) can be represented by a one-parameter family of maps[6] $\phi_t : U \to \mathbb{R}^m$ satisfying

$$\frac{d}{dt}[\phi(t, x)]\bigg|_{t=\tau} = f[\phi(\tau, x)]$$

for all $x \in U$ and for all $\tau \in I$ for which the solution is defined.

The family of maps $\phi_t(x) = \phi(t, x)$ is called the flow (or the flow map) generated by the vector field f. If f is continuously differentiable (that is, if all the functions in the vector are continuously differentiable), then for any point x_0 in the domain U there exists a $\delta(x_0) > 0$ such that the solution $\phi(t, x_0)$ through that point exists and is unique for $|t| < \delta$. The existence and uniqueness result is *local* in time in the sense that δ need not extend to (plus or minus) infinity and certain vector fields have solutions that 'explode' in finite time (see exercise 1.8(c) at the end of the chapter).

When the solution of a system of differential equations $\dot{x} = f(x)$ is not defined for all time, a new system $\dot{x} = g(x)$ can be determined which has the same forward and backward orbits *in the state space* and such that each orbit is defined for all time. If $\psi(t, x)$ is the flow generated by the vector field g, the relation between ψ and the flow ϕ generated by f is the following:

$$\psi(t, x) = \phi[\tau(t, x), x] \qquad x \in U$$

and

$$\tau : \mathbb{R} \times U \to \mathbb{R}$$

is a time-reparametrisation function monotonically increasing in t for all $x \in U$.

EXAMPLE Suppose we have a system

$$\dot{x} = f(x) \tag{1.16}$$

with $f : \mathbb{R}^m \to \mathbb{R}^m$, a continuously differentiable function with flow $\phi(t, x)$ defined on a maximal time interval $-\infty < a < 0 < b < +\infty$. Then the

---

[6]The terms map or mapping indicate a function. In this case, we speak of $y = f(x)$ as the **image** of x under the map f. If f is invertible, we can define the inverse function f^{-1}, that is, the function satisfying $f^{-1}[f(x)] = x$ for all x in the domain of f and $f[f^{-1}(y)] = y$ for all y in the domain of f^{-1}. Even if f is not invertible, the notation $f^{-1}(y)$ makes sense: it is the **set of pre-images** of y, that is, all points x such that $f(x) = y$. The terms map, mapping are especially common in the theory of dynamical systems where iterates of a map are used to describe the evolution of a variable in time.

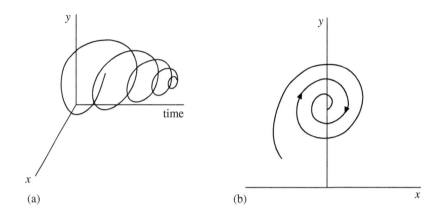

Fig. 1.5 A damped oscillator in $\mathbb{R}^2$: (a) space of motions; (b) state space

differential equation $\dot{x} = g(x)$ with $g(x) : \mathbb{R}^m \to \mathbb{R}^m$ and

$$g(x) = \frac{f(x)}{1 + \|f(x)\|}$$

(where $\| \cdot \|$ denotes Euclidean norm), defines a dynamical system whose forward and backward orbits are the same as those of (1.16) but whose solutions are defined for all time.[7]

The set of points $\{\phi(t, x_0) \mid t \in I\}$ defines an **orbit** of (1.14), starting from a given point x_0. It is a solution curve in the state space, parametrised by time. The set $\{[t, \phi(t, x_0)] \mid t \in I\}$ is a **trajectory** of (1.14) and it evolves in the space of motions. However, in applications, the terms orbit and trajectory are often used as synonyms. A simple example of a trajectory in the space of motions $\mathbb{R} \times \mathbb{R}^2$ and the corresponding orbit in the state space $\mathbb{R}^2$ is given in figure 1.5. Clearly the orbit is obtained by projecting the trajectory onto the state space.

The flows generated by vector fields form a very important subset of a more general class of maps, characterised by the following definition.

definition 1.1 *A **flow** is a map $\phi : I \subset \mathbb{R} \times X \to X$ where X is a metric space, that is, a space endowed with a distance function, and ϕ has the following properties*

(a) $\phi(0, x) = x$ *for every* $x \in X$ *(identity axiom);*
(b) $\phi(t + s, x) = \phi[s, \phi(t, x)] = \phi[t, \phi(s, x)] = \phi(s + t, x)$, *that is, time-translated solutions remain solutions;*

[7] For details see Bhatia and Szegö (1970), p. 78; Robinson (1999), pp. 146–7.

(c) for fixed t, ϕ_t is a homeomorphism on X.

Alternatively, and equivalently, a flow may be defined as a one-parameter family of maps $\phi_t : X \to X$ such that the properties *(a)–(c)* above hold for all $t, s \in \mathbb{R}$.

remark 1.2 A **distance** on a space X (or, a **metric** on X) is a function $X \times X \to \mathbb{R}^+$ satisfying the following properties for all $x, y \in X$:

(1) $d(x, y) \geq 0$ and $d(x, y) = 0$ if and only if $x = y$;
(2) $d(x, y) = d(y, x)$ (symmetry);
(3) $d(x, y) \leq d(x, z) + d(z, y)$ (triangle inequality).

Notice that there also exist notions of distance which are perfectly meaningful but do not satisfy the definition above and therefore do not define a metric, for example:

the distance between a point and a set A;

$$d(x, A) = \inf_{y \in A} d(x, y).$$

the distance between two sets A and B

$$d(A, B) = \inf_{x \in A} \inf_{y \in B} d(x, y).$$

Neither of these cases satisfies property (1) in remark 1.2. However, there exists a 'true' distance between two sets which is a metric in the space of nonempty, compact sets, i.e., the Hausdorff distance.[8]

In this book we are mainly concerned with applications for which ϕ is a flow generated by a system of differential equations and the state space is an Euclidean space or, sometimes, a manifold. However, some concepts and results in later chapters of the book will be formulated more generally in terms of flows on metric spaces.

Consider now a system of nonautonomous differential equations such that

$$\dot{x} = f(t, x) \tag{1.17}$$

where $f : \mathbb{R} \times U \to \mathbb{R}^m$, and assume that a unique solution exists for all $(t_0, x_0) \in \mathbb{R} \times U$. Then we can represent solutions of (1.17) by means of a flow $\phi : \mathbb{R} \times X \to X$, where $X \subset (\mathbb{R} \times \mathbb{R}^m)$. This suggests that a

---

[8] See, for example, Edgar (1990), pp. 65–6.

nonautonomous system $\dot{x} = f(t, x)$ can be transformed into an equivalent autonomous system by introducing an arbitrary variable $\theta = t$ and writing

$$\dot{\theta} = 1$$
$$\dot{x} = f(\theta, x). \tag{1.18}$$

Notice that, by definition, the extended autonomous system (1.18) has no equilibrium point in X. However, if the original, nonautonomous system (1.17) has a *uniformly* stable (*uniformly*, asymptotically stable) equilibrium point, then for the extended autonomous system (1.17), the t-axis is a stable (asymptotically stable) invariant set. The precise meaning of (asymptotic, uniform) stability will be discussed in chapters 3 and 4.

Solutions of system (1.14) can be written in either the simpler form $x(t)$, $x : I \rightarrow \mathbb{R}^m$, or $\phi_t(x) : U \rightarrow \mathbb{R}^m$, or again $\phi(t, x)$, $\phi : I \times U \rightarrow \mathbb{R}^m$, depending on what aspect of solutions one wants to emphasise. The notation $\phi_t(x)$ is especially suitable for discussing discrete-time maps derived from continuous-time systems.

If time t is allowed to take only uniformly sampled, discrete values, separated by a fixed interval τ, from a continuous-time flow we can derive a discrete-time map (a difference equation)

$$x_{n+\tau} = G(x_n) \tag{1.19}$$

where $G = \phi_\tau$. Certain properties of continuous-time dynamical systems are preserved by this transformation and can be studied by considering the discrete-time systems derived from them. If the unit of measure of time is chosen so that $\tau = 1$, we have the canonical form

$$x_{n+1} = G(x_n). \tag{1.20}$$

Let the symbol $\circ$ denote the composition of functions, so that, $f \circ g(x)$ means $f[g(x)]$. Then we write

$$x_n = G(x_{n-1}) = G \circ G(x_{n-2}) = \ldots = G \circ G \circ \ldots \circ G(x_0) = G^n(x_0)$$

where G^n is the composition of G with itself n times, or the nth iteration of G, with $n \in \mathbb{Z}^+$. If G is invertible and G^{-1} is a well defined function, G^n with $n \in \mathbb{Z}^-$ denotes the nth iterate of G^{-1}. (Note that $G^n(x)$ is *not* the nth power of $G(x)$.) Thus, iterates of the map G (or G^{-1}) can be used to determine the value of the variable x at time n, when the initial condition x_0 is fixed.[9]

---

[9] For autonomous difference equations whose solutions do not depend on the choice of the initial time, in a manner analogous to our practice for autonomous differential equations, we take the initial time as zero.

remark 1.3 There exists another way of deriving a discrete-time map from a continuous-time dynamical system, called **Poincaré map**, which describes the sequence of positions of a system generated by the intersections of an orbit in continuous time and a given space with a lower dimension, called **surface of section**. Clearly, in this case the time intervals between different pairs of states of the systems need not be equal. Poincaré maps are a powerful method of investigation of dynamical systems and we shall make some use of them in chapter 4, when we discuss periodic solutions and in chapters 7 and 8.

Of course, there exist problems that are conceived from the beginning as discrete dynamical systems (difference equations). In fact, there are difference equations that cannot be derived from differential equations. In particular, this is true of *noninvertible maps* which have been extensively used in recent years in the study of dynamical problems in many applied fields. Intuitively, the reason why a noninvertible map cannot be a flow map (derived from a differential equation as explained above) is that such a map uniquely determines the dynamics in one time direction only whereas, under standard assumptions, solutions of a differential equation always determine the dynamics in both directions uniquely.

remark 1.4 Orbits of differential equations are continuous *curves*, while orbits of maps are *discrete sets of points*. This has a number of important consequences, the most important of which can be appreciated intuitively. If the solution of an autonomous system of differential equations through a point is unique, two solution curves cannot intersect one another in the state space. It follows that, for continuous-time dynamical systems of dimension one and two, the orbit structure must be drastically constrained. In the former, simpler case, we can only have fixed points and orbits leading to (or away from) them; in the two-dimensional case, nothing more complex than periodic orbits can occur. For maps the situation is different. It remains true that the orbit starting from a given point in space is uniquely determined in the direction defined by the map. However, since discrete-time orbits, so to speak, can 'jump around', even simple, one-dimensional nonlinear maps can generate very complicated orbits, as we shall see in the following chapters.

Generalising the simple examples discussed in sections 1.2 and 1.3 above, the stationary, equilibrium solutions of multi-dimensional dynamical systems in both continuous and discrete time can be identified by solving systems of equations.

In the former case, setting $\dot{x} = 0$ in (1.14) the set of **equilibrium** or

fixed points is defined by

$$E = \{\bar{x} | f(\bar{x}) = 0\}$$

that is, the set of values of x such that its rate of change in time is nil.

Analogously, in the discrete-time case,

$$x_{n+1} = G(x_n)$$

we have

$$E = \{\bar{x} | \bar{x} - G(\bar{x}) = 0\}$$

that is, the set of values of x that are mapped to themselves by G. Because the functions f and G are generally nonlinear, there are no ready-made procedures to find the equilibrium solutions exactly, although geometrical and numerical techniques often give us all the qualitative information we need. Notice that linear systems typically have a unique solution, whereas nonlinear systems typically have either no solutions, or a finite number of them. It follows that only nonlinear systems may describe the interesting phenomenon of (finite) multiple equilibria.

For a system of autonomous, differential equations like (1.14), a general solution $\phi(t, x)$ can seldom be written in a closed form, i.e., as a combination of known elementary functions (powers, exponentials, logarithms, sines, cosines, etc.). Unfortunately, closed-form solutions are available only for special cases, namely: systems of linear differential equations; one-dimensional differential equations (i.e., those for which $m = 1$); certain rather special classes of nonlinear differential equations of order greater than one (or systems of equations with $m > 1$). The generality of nonlinear systems which are studied in applications escapes full analytical investigation, that is to say, an exact mathematical description of solution orbits cannot be found. Analogous difficulties arise when dynamical systems are represented by means of nonlinear maps. In this case, too, closed-form solutions are generally available only for linear systems.

The importance of this point should not be exaggerated. On the one hand, even when a closed-form solution exists, it may not be very useful. A handbook of mathematical formulae will typically have a hundred pages of integrals for specific functions, so that a given nonlinear model may indeed have a solution. However, that solution may not provide much intuition, nor much information if the solution function is not a common, well known function. On the other hand, in many practical cases we are not especially interested in determining (or approximating) exact individual solutions, but we want to establish the qualitative properties of an ensemble of orbits starting from certain practically relevant sets of initial conditions. These properties can often be investigated effectively by a combination of mathematical, geometrical, statistical and numerical methods. Much of what follows is dedicated precisely to the study of some of those methods.

Before turning to this goal, however, we review in chapter 2 the class of dynamical systems which is best understood: linear systems. Dynamical linear systems in both continuous and discrete time are not terribly interesting *per se* because their behaviour is morphologically rather limited and they cannot be used effectively to represent cyclical or complex dynamics. However, linear theory is an extremely useful tool in the analysis of nonlinear systems. For example, it can be employed to investigate qualitatively their **local** behaviour, e.g., their behaviour in a neighbourhood of a single point or of a periodic orbit. This is particularly important in stability analysis (chapters 3 and 4) and in the study of (local) bifurcations (chapter 5).

Exercises

1.1 Consider the discrete-time partial equilibrium model summarised in (1.6) given the parameter values $a = 10$, $b = 0.2$, $m = 2$, $s = 0.1$. Write the general solution given the initial values $p_0 = 20$ and $p_0 = 100$. Calculate the values for the price at time periods 0, 1, 2, 4, 10, 100 starting from each of the above initial values and sketch the trajectories for time periods 0–10.

1.2 State a parameter configuration for the discrete-time partial equilibrium model that implies $\beta < 0$. Describe the dynamics implied by that choice. Using these parameter values and $a = 10$, $m = 2$, sketch the dynamics in the space (p_n, p_{n+1}). Draw the bisector line and from the chosen initial condition, iterate 3 or 4 times. Show the direction of movement with arrows.

1.3 If we define the parameters as in exercise 1.1 ($b = 0.2$, $s = 0.1$, $a = 10$, $m = 2$), the continuous-time, partial equilibrium model of (1.11) gives the constant exponent of the solution as $b + s = 0.3$. Let this be case 1. If $s = 0.6$, $b + s = 0.8$. Let this be case 2. Calculate the solution values for case 1 and case 2 at periods $t = 0, 1, 2, 4.67, 10, 100$ starting from the initial condition $p_0 = 20$. Comment on the speed of the adjustment process. Note the different integer values of t for which equilibrium in Case 2 is approximated using a precision of 1 decimal point, 2 decimal points, 3 decimal points.

1.4 Suppose that the good under consideration is a 'Giffen' good (for which $dD/dp > 0$ and therefore $b < 0$). It is unlikely, but possible that $b + s < 0$. Sketch the differential equation (1.10) under that hypothesis in the $(p, \dot{p})$ plane, note the equilibrium point and comment on the adjustment process.

1.5   Convert these higher-order differential equations to systems of first-order differential equations and write the resulting systems in matrix form:

(a) $\ddot{x} + x = 1$
(b) $\frac{d^3 x}{dt^3} + 0.4\ddot{x} - 2x = 0$
(c) $\frac{d^4 x}{dt^4} + 4\ddot{x} - 0.5\dot{x} - x = 11$.

1.6   Convert the following higher-order system of differential equations into a system of first-order differential equations

$$\ddot{x} + x = 1$$
$$\ddot{y} - \dot{y} - y = -1.$$

1.7   Higher-order difference equations and systems can also be converted to first-order systems using auxiliary variables. A kth-order equation $x_{n+k} = G(x_{n+k-1}, \ldots, x_n)$ can be converted by setting

$$x_n = z_n^{(1)}$$
$$z_{n+1}^{(1)} = x_{n+1} = z_n^{(2)}$$
$$z_{n+1}^{(2)} = x_{n+2} = z_n^{(3)}$$
$$\vdots \qquad \vdots \qquad \vdots$$
$$z_{n+1}^{(k)} = x_{n+k} = G(x_{n+k-1}, \ldots, x_n) = G(z_n^{(k)}, \ldots, z_n^{(1)}).$$

Convert the following difference equations into systems of first-order difference equations and write them in matrix form

(a) $x_{n+2} - a x_{n+1} + b x_n = 1$
(b) $0.5 x_{n+3} + 2 x_{n+1} - 0.1 x_n = 2$.

1.8   Use integration techniques to find exact solutions to the following differential equations and sketch trajectories where possible, assuming an initial value of $x(0) = 1$

(a) $\dot{x} = 2x$
(b) $\dot{x} = \frac{1}{x^2}$
(c) $\dot{x} = x^2$.

1.9   Use the technique described in the example in section 1.4 to find a function g, defined over all time and such that $\dot{x} = g(x)$ has the same backward and forward orbits in the state space as $\dot{x} = x^2$.

1.10   Write the exact solution of the following differential equation (*Hint:* rewrite the equation as $dx/dt = \mu x(1-x)$ and integrate, separating variables) and discuss the dynamics of x

$$\dot{x} = \mu x(1-x) \qquad x \in [0,1].$$

2

Review of linear systems

The theory of linear dynamical systems is a very well-developed area of research and even an introductory presentation would easily fill a large book in itself. Many excellent mathematical treatments of the matter exist and we refer the reader to them for more details. Our objective in what follows is to provide a review of the theory, concentrating on those aspects which are necessary to understand our discussion of nonlinear dynamical systems. We consider the general multi-dimensional case first and then, to fix certain basic ideas, we discuss in detail simple mathematical examples in the plane, as well as an extended version of the economic partial equilibrium model.

2.1 Introduction

Broadly speaking, we say that a phenomenon represented by a stimulus-response mechanism is **linear** if, to a given change in the intensity of the stimulus, there corresponds a proportional change in the response. Thus, postulating that saving is a linear function of income implies that a doubling of the level of income doubles the amount of saving as well.

As concerns dynamical systems, we say that systems such as

$$\frac{dx}{dt} = \dot{x} = f(x) \qquad x \in \mathbb{R}^m \quad t \in \mathbb{R} \tag{2.1}$$

or

$$x_{n+1} = G(x_n) \qquad x \in \mathbb{R}^m \quad n \in \mathbb{Z} \tag{2.2}$$

are linear if the vector-valued functions $f(x)$ and $G(x_n)$ are linear according to the following definition:

definition 2.1 *A function $f : \mathbb{R}^m \to \mathbb{R}^m$ is **linear** if $f(\alpha v + \beta w) = \alpha f(v) + \beta f(w)$, for any $\alpha, \beta \in \mathbb{R}$ and $v, w \in \mathbb{R}^m$.*

In chapter 1 we provided a preliminary discussion of some simple dynamical systems and said that, since they were linear, the sum of solutions was also a solution. We can now generalise that idea and make it more precise, beginning with systems in continuous time. Linear systems of differential equations satisfy the **superposition principle**, defined as follows: if $\phi_1(t, x)$ and $\phi_2(t, x)$ are any two linearly independent[1] solutions of system (2.1) then

$$S(t, x) = \alpha\phi_1(t, x) + \beta\phi_2(t, x) \tag{2.3}$$

is also a solution for any $\alpha, \beta \in \mathbb{R}$. It can be verified that the superposition principle is valid if, and only if, the (vector-valued) function f of (2.1) is linear. If $\phi_1(t, x)$ and $\phi_2(t, x)$ are solutions to (2.1) then, respectively:

$$\frac{d}{dt}[\phi_1(t, x)]\Big|_{t=\tau} = f[\phi_1(\tau, x)]$$

$$\frac{d}{dt}[\phi_2(t, x)]\Big|_{t=\tau} = f[\phi_2(\tau, x)].$$

On the other hand, in order for (2.3) to be a solution, we must have

$$\frac{d}{dt}[S(t, x)]\Big|_{t=\tau} = f[S(\tau, x)]$$

or

$$\alpha f[\phi_1(\tau, x)] + \beta f[\phi_2(\tau, x)] = f[\alpha\phi_1(\tau, x) + \beta\phi_2(\tau, x)]$$

which holds if, and only if, f is linear.

An entirely analogous argument can be developed for the discrete-time system (2.2) and we leave it to the reader as an exercise.

When f and G are vectors of linear functions, their elements can be written as products of constants and the variables, namely,

$$f_i(x) = f_{i1}x_1 + f_{i2}x_2 + \cdots + f_{im}x_m \qquad (i = 1, 2, \ldots, m)$$

and

$$G_i(x) = G_{i1}x_1 + G_{i2}x_2 + \cdots + G_{im}x_m \qquad (i = 1, 2, \ldots, m),$$

respectively, where f_i and G_i denote the ith element of f and G (called **coordinate function**), respectively, and f_{ij}, G_{ij} $(i, j = 1, 2, \ldots, m)$ are constants.

---

[1] We say that $\phi_1(t, x)$ and $\phi_2(t, x)$ are **linearly independent** if, and only if, $\alpha\phi_1(t, x) + \beta\phi_2(t, x) = 0$ for all x and all t implies that $\alpha = \beta = 0$.

Then systems (2.1) and (2.2) can be put in a compact, matrix form, respectively, as

$$\dot{x} = Ax \qquad x \in \mathbb{R}^m \tag{2.4}$$

and

$$x_{n+1} = Bx_n \qquad x \in \mathbb{R}^m \tag{2.5}$$

where A and B are $m \times m$ matrices of constants, also called the coefficient matrices, with typical elements f_{ij} and G_{ij}, respectively.

remark 2.1 Consider the equation

$$\dot{x} = f(x) = a + bx \qquad x \in \mathbb{R} \tag{2.6}$$

where a and b are scalar constants. If we try definition 2.1 on the function $f(x)$, letting x take on the values v and w, we have

$$f(\alpha v + \beta w) = a + b\alpha v + b\beta w.$$

But

$$\alpha f(v) + \beta f(w) = \alpha a + \alpha bv + \beta a + \beta bw$$

and, for $a \neq 0$, these are equal if, and only if, $\alpha + \beta = 1$. Therefore the requirements of definition 2.1 are not satisfied and the superposition principle does not hold for (2.6). Functions like f are linear plus a translation and are called **affine**.

Dynamical systems characterised by affine functions though, strictly speaking, nonlinear, can easily be transformed into linear systems by translating the system without affecting its qualitative dynamics.

We show this for the general, continuous-time case. Consider the system

$$\dot{y} = Ay + c \tag{2.7}$$

where $y \in \mathbb{R}^m$, A is an $m \times m$ matrix and $c \in \mathbb{R}^m$ is a vector of constants.

We can set

$$x = y + k$$

so that

$$\dot{x} = \dot{y} = Ay + c = Ax - Ak + c.$$

If A is **nonsingular** (the determinant, $\det(A)$, is nonzero and therefore there are no zero eigenvalues), A^{-1} is well defined and we can set $k = A^{-1}c$ whence $\dot{x} = Ax$ which is linear and has the same dynamics as (2.7). Notice that the assumption that A is nonsingular is the same condition required for determining the fixed (equilibrium) points of (2.7). Recall that a point $\bar{y}$ is a fixed or equilibrium point for (2.7) if $\dot{y} = A\bar{y} + c = 0$, whence

$$\bar{y} = -A^{-1}c.$$

If A is singular, that is, if A has a zero eigenvalue, the inverse matrix A^{-1} and, therefore, the equilibrium are indeterminate.

The discrete-time case is entirely analogous. Consider the system

$$y_{n+1} = By_n + c \qquad (2.8)$$

where $y_n \in \mathbb{R}^m$, B is a $m \times m$ matrix and $c \in \mathbb{R}^m$ is a vector of constants. We set

$$x_n = y_n + k$$

so that

$$x_{n+1} = y_{n+1} + k = By_n + c + k = Bx_n - Bk + c + k.$$

If we set $k = -[I - B]^{-1}c$ (where I denotes the identity matrix), we have $x_{n+1} = Bx_n$, a linear system of difference equations with the same dynamics as (2.8). Once again k is defined if the matrix $[I - B]$ is nonsingular, i.e., if none of the eigenvalues of the matrix B is equal to one. This is also the condition for determinateness of the equilibrium $\bar{y} = [I - B]^{-1}c$ of system (2.8).

2.2 General solutions of linear systems in continuous time

In this section we discuss the form of the solutions to the general system of linear differential equations (2.4)

$$\dot{x} = Ax \qquad x \in \mathbb{R}^m.$$

First, notice that the function $x(t) = 0$ solves (2.4) trivially. This special solution is called the **equilibrium solution** because if $x = 0$, $\dot{x} = 0$ as well. That is, a system starting at equilibrium stays there forever. Notice that if A is nonsingular, $x = 0$ is the only equilibrium for linear systems like (2.4). Unless we indicate otherwise, we assume that this is the case.

Let us now try to determine the nontrivial solutions. In chapter 1, we learned that, in the simple one-dimensional case, the solution has the exponential form $ke^{\lambda t}$, k and λ being two scalar constants. It is natural to wonder whether this result can be generalised, that is, whether there exist (real or complex) constant vectors u and (real or complex) constants λ such that

$$x(t) = e^{\lambda t} u \qquad (2.9)$$

is a solution of (2.4). Differentiating (2.9) with respect to time, and substituting into (2.4), we obtain

$$\lambda e^{\lambda t} u = A e^{\lambda t} u$$

which, for $e^{\lambda t} \neq 0$, implies

$$Au = \lambda u \quad \text{or} \quad (A - \lambda I)u = 0 \qquad (2.10)$$

where 0 is an m-dimensional null vector. A nonzero vector u satisfying (2.10) is called an **eigenvector** of matrix A associated with the **eigenvalue** λ. Equation (2.10) has a nontrivial solution $u \neq 0$ if and only if

$$\det(A - \lambda I) = 0. \qquad (2.11)$$

Equation (2.11) is called the **characteristic equation** of system (2.4) and can be expressed in terms of a polynomial in λ, thus

$$\det(A - \lambda I) = \mathcal{P}(\lambda) = \lambda^m + k_1 \lambda^{m-1} + \cdots + k_{m-1}\lambda + k_m = 0 \qquad (2.12)$$

and $\mathcal{P}(\lambda)$ is called the **characteristic polynomial**. The answer to our question is, therefore, yes, (2.9) is indeed a solution of (2.4) if λ is an eigenvalue and u is the associated eigenvector of A. Thus, we have reduced the problem of solving a functional equation (2.4) to the algebraic problem of finding the roots of the polynomial (2.12) and solving the system of equations (2.10) which, for given λ, is linear. Notice that for each λ, if u is an eigenvector, so is cu for $c \neq 0$.

remark 2.2 The coefficients k_i of (2.12) depend on the elements of the matrix A and, in principle, can be determined in terms of those elements by means of rather complicated formulae. However, the first and the last, namely k_1 and k_m, can be related in a particularly simple manner, to the matrix A and its eigenvalues as follows:

$$k_1 = -\operatorname{tr}(A) = -\sum_{i=1}^{m} \lambda_i$$

$$k_m = (-1)^m \det(A) = (-1)^m \prod_{i=1}^{m} \lambda_i$$

where $\operatorname{tr}(A)$ denotes the trace of A, that is, the sum of the elements on the main diagonal and, again, $\det(A)$ denotes the determinant of A.

Equation (2.12) has m roots which may be real or complex and some of them may be repeated. In order to render the discussion more fluid we give general results under the assumption that eigenvalues are distinct, followed by some comments on the more complicated case of repeated eigenvalues. As regards complex roots, if the elements of matrix A are real, complex eigenvalues (and eigenvectors) always appear in conjugate pairs.

Suppose there exist n ($0 \leq n \leq m$) real, distinct eigenvalues λ_i of the matrix A with corresponding n real, linearly independent eigenvectors u_i, and $p = (m - n)/2$ pairs of distinct complex conjugate[2] eigenvalues $(\lambda_j, \bar{\lambda}_j)$ with corresponding eigenvectors $(u_j, \bar{u}_j)$ where ($0 \leq p \leq m/2$). Then, in the real case, there are n linearly independent solutions of (2.4) with the form

$$x_i(t) = e^{\lambda_i t} u_i \qquad (i = 1, 2, \ldots, n). \qquad (2.13)$$

In the complex case, each pair of eigenvalues and associated pair of eigenvectors can be written, respectively, as

$$(\lambda_j, \lambda_{j+1}) = (\lambda_j, \bar{\lambda}_j) = (\alpha_j + i\beta_j, \alpha_j - i\beta_j)$$

where α, β are real scalars, and

$$(u_j, u_{j+1}) = (u_j, \bar{u}_j) = (a_j + ib_j, a_j - ib_j)$$

where a_j and b_j are m-dimensional vectors and, $i^2 = -1$. The two corresponding *real* solutions of (2.4) are the vectors

$$
\begin{aligned}
x_j(t) &= \frac{1}{2}\left(e^{\lambda_j t} u_j + e^{\bar{\lambda}_j t} \bar{u}_j\right) \\
&= e^{\alpha_j t}\left[a_j \cos(\beta_j t) - b_j \sin(\beta_j t)\right] \\
x_{j+1}(t) &= -\frac{i}{2}\left(e^{\lambda_j t} u_j - e^{\bar{\lambda}_j t} \bar{u}_j\right) \\
&= e^{\alpha_j t}\left[a_j \sin(\beta_j t) + b_j \cos(\beta_j t)\right]
\end{aligned}
\qquad (2.14)
$$

where we have employed Euler's formula $e^{\pm i\theta} = \cos\theta \pm i\sin\theta$ to write

$$e^{(\alpha \pm i\beta)t} = e^{\alpha t}\left[\cos(\beta t) \pm i\sin(\beta t)\right].$$

Equations (2.14) can be further simplified by transforming the real and imaginary parts of each element of the complex vectors $(u_j, \bar{u}_j)$ into the

---

[2]If $z = \alpha + i\beta$ is an arbitrary complex number, then the complex number $\bar{z} = \alpha - i\beta$ is the **complex conjugate** of z. The same notation is used for vectors.

polar coordinates

$$a_j^{(l)} = C_j^{(l)} \cos(\phi_j^{(l)})$$
$$b_j^{(l)} = C_j^{(l)} \sin(\phi_j^{(l)}) \qquad l = 1, 2, \ldots, m$$

where $a_j^{(l)}$ and $b_j^{(l)}$ are the lth elements of a_j and b_j, respectively.

Making use of the trigonometric identities

$$\sin(x \pm y) = \sin x \cos y \pm \cos x \sin y$$
$$\cos(x \pm y) = \cos x \cos y \mp \sin x \sin y$$

(2.15)

the solutions (2.14) of (2.4) can be written as m-dimensional vectors whose lth elements have the form

$$x_j^{(l)}(t) = C_j^{(l)} e^{\alpha_j t} \cos(\beta_j t + \phi_j^{(l)})$$
$$x_{j+1}^{(l)}(t) = C_j^{(l)} e^{\alpha_j t} \sin(\beta_j t + \phi_j^{(l)})$$

(2.16)

Applying the superposition principle, we can now write the general solution of (2.4) as a function of time and m arbitrary constants, namely:

$$x(t) = c_1 x_1(t) + c_2 x_2(t) + \cdots + c_m x_m(t)$$

(2.17)

where $x_i(t)$ $(i = 1, \ldots, m)$ are the individual solutions defined in (2.13), (2.14) (or (2.16)) and the m constants c_i are determined by the initial conditions. It is possible to show that formula (2.17) gives all of the possible solutions to (2.4).

Simple inspection of (2.13) and (2.14) suggests that if *any* of the eigenvalues has a positive real part, solutions initiating from a generic point (one for which $c_i \neq 0 \; \forall_i$) diverge, i.e., variables tend to $+$ or $-\infty$ as time tends to plus infinity. On the contrary, if *all* eigenvalues have negative real parts, solutions asymptotically tend to zero, that is, to the unique equilibrium point. Anticipating our systematic discussion of stability in chapter 3, we shall say that in the latter case, the equilibrium is **asymptotically stable**, in the former case the equilibrium is **unstable**. The situation is more complicated if there are no eigenvalues with positive real parts but some eigenvalues have zero real parts. If there is only one such eigenvalue, the system has an attenuated form of stability — it does not converge to equilibrium, but it does not diverge either. (If one of the eigenvalues is zero, the equilibrium need not be unique.)

Consider now, the case of a real eigenvalue λ_k and set the initial conditions so that $c_k \neq 0$, $c_i = 0 \; \forall_{i \neq k}$. From (2.13) and (2.17) we have

$$x(0) = c_k x_k(0) = c_k u_k.$$

(2.18)

Thus, we have chosen initial conditions so as to position the system, at time zero, on the eigenvector associated with λ_k. Using (2.13), (2.17) and (2.18) we deduce that the solution for $|t| > 0$ in this case is given by

$$x(t) = c_k x_k(t) = e^{\lambda_k t} c_k u_k = e^{\lambda_k t} x(0).$$

Then, if the initial conditions are on the eigenvector, the system either approaches or moves away from the equilibrium point (depending on the sign of the eigenvalue), *along that vector*. In other words, each real eigenvector u_k defines a direction of motion in the state space that is preserved by the matrix A. In this case, we say that eigenvectors are invariant sets. Broadly speaking, a certain region of the state space is said to be **invariant** with respect to the action of a continuous- or discrete-time dynamical system, if an orbit starting in the set remains there forever (unless disturbed by exogenous shocks). The speed of the motion along the eigenvector u_k is given by $\dot{x}(t) = \lambda_k e^{\lambda_k t} x(0) = \lambda_k x(t)$, that is, the speed is a constant proportion of the position of x.

The complex case is more difficult to analyse, but it is possible to show that the plane S spanned (or generated) by the two linearly independent real vectors associated with a pair of complex conjugate eigenvectors $(u_j, \bar{u}_j)$,

$$\frac{1}{2}(u_j + \bar{u}_j) = a_j$$

$$-\frac{i}{2}(u_j - \bar{u}_j) = b_j$$

(2.19)

is invariant. (Vectors a_j and b_j are the real and imaginary parts of $(u_j, \bar{u}_j)$.) If we choose the initial conditions on S so that the coefficients c_i of the general solution are all zero except the two corresponding to the pair of complex eigenvalues $(\lambda_j, \bar{\lambda}_j)$, the orbits of the system remain on the plane S forever.

Again, assuming that the matrix A has m distinct eigenvalues, we divide the eigenvectors (or, in the complex case, the vectors equal to the real and imaginary parts of them) into three groups, according to whether the corresponding eigenvalues have negative, positive or zero real parts. Then the subsets of the state space spanned (or generated) by each group of vectors are known as the **stable**, **unstable** and **centre eigenspaces**, respectively, and denoted by E_s, E_u and E_c. If m_s is the number of eigenvalues with negative real parts, m_u the number with positive real parts and m_c the number with zero real parts, the dimension of the stable, unstable and centre eigenspaces are m_s, m_u and m_c, respectively, and $m = m_s + m_u + m_c$. The subspaces E_s, E_u and E_c are invariant. The notion of stable, unstable and centre eigenspaces can be generalised to the case of repeated eigenvalues, but we do not discuss the question here.

There exists a nice, compact way of writing the general solution of (2.4). Let $X(t)$ be the matrix whose m columns are the m solutions defined by (2.13) and (2.14), or (2.16), that is,

$$X(t) = [x_1(t) \ x_2(t) \ \dots \ x_m(t)].$$

$X(t)$ is called the **fundamental matrix solution** of $\dot{x} = Ax$ and it has some interesting properties, in particular:

(i) $X(t) = e^{tA}X(0)$, where e^{tA} is a converging, infinite sum of $(m \times m)$ matrices defined as

$$e^{tA} = \left(I + tA + \frac{t^2}{2!}A^2 + \dots + \frac{t^n}{n!}A^n + \dots \right);$$

(ii) $\dot{X}(t) = AX(t)$, that is, $X(t)$ satisfies (2.4) as a matrix equation;
(iii) if the solutions $x_i(t)$ are linearly independent, $X(t)$ is nonsingular for all t.

remark 2.3 The reader can verify, looking at (2.13) and (2.14), that the columns of $X(0)$ are of two types:

(1) for a solution of type (2.13) (real eigenvalues and eigenvectors), the columns are equal to u_i;
(2) for each pair of solutions of type (2.14) with $(u_j, u_{j+1}) = (a_j + ib_j, a_j - ib_j)$ (complex conjugate eigenvalues and eigenvectors), the corresponding two columns are equal to the real vectors a_j and b_j, respectively.

From (2.17) we have

$$x(t) = X(t)c$$

where $c = (c_1, c_2, \dots, c_m)^T$ and therefore, from property (i) of matrix X, the solutions of (2.4) can be written as

$$x(t) = e^{tA}x(0). \tag{2.20}$$

We can verify that (2.20) is actually a solution of (2.4) by differentiating with respect to time and obtaining

$$\dot{x}(t) = Ae^{tA}x(0) = Ax(t)$$

which is (2.4).

When there are m distinct, real eigenvalues (and m linearly independent real eigenvectors), the fundamental matrix solution $X(t)$ can be used to transform the matrix A into a diagonal matrix having the same eigenvalues as A. The reader can verify that

$$AX(t) = X(t)\Lambda \qquad \Lambda = \begin{pmatrix} \lambda_1 & \cdots & 0 \\ \vdots & \ddots & \vdots \\ 0 & \cdots & \lambda_m \end{pmatrix}.$$

Because $X(t)$ is nonsingular we have

$$X(t)^{-1}AX(t) = \Lambda.$$

When there are distinct, complex eigenvalues the transformation, as usual, is more complicated and the resulting operation is a *block diagonalisation*. Consider a simple case in which A is a (2×2) matrix with two complex conjugate eigenvalues $\lambda = \alpha + i\beta$, $\bar{\lambda} = \alpha - i\beta$, and two corresponding eigenvectors $u = a + ib$ and $\bar{u} = a - ib$. Recalling that $\text{tr}(A) = \lambda + \bar{\lambda} = 2\alpha$ and $\det(A) = \lambda\bar{\lambda} = \alpha^2 + \beta^2$, the characteristic polynomial in this case can be written as

$$\lambda^2 - 2\alpha\lambda + (\alpha^2 + \beta^2) = 0.$$

Making use of a theorem by Cayley and Hamilton stating that if $\mathcal{P}(\lambda)$ is the characteristic polynomial of a matrix A, then the matrix polynomial $\mathcal{P}(A) = 0$ (where 0 is a matrix of the same order as A in which all elements are null), we can write

$$A^2 - 2\alpha A + (\alpha^2 + \beta^2)I = 0$$

or, factoring,

$$(A - \lambda I)(A - \bar{\lambda}I) = (A - \alpha I)^2 + \beta^2 I = 0. \qquad (2.21)$$

If x is any nonzero real $(n \times 1)$ vector and $y = (A - \alpha I)x/\beta$ (x, y are linearly independent if $\beta \neq 0$), we have

$$Ax = \beta y + \alpha x. \qquad (2.22)$$

Pre-multiplying (2.22) by $(A - \alpha I)$ and making use of (2.21) we have

$$Ay = \alpha y - \beta x.$$

Then defining

$$P = \begin{pmatrix} x^{(1)} & y^{(1)} \\ x^{(2)} & y^{(2)} \end{pmatrix}$$

where $x^{(i)}$, $y^{(i)}$ denote the ith elements of the vectors x,y, respectively, we have

$$AP = P \begin{pmatrix} \alpha & -\beta \\ \beta & \alpha \end{pmatrix}$$

or

$$P^{-1}AP = \begin{pmatrix} \alpha & -\beta \\ \beta & \alpha \end{pmatrix}. \tag{2.23}$$

The matrix on the RHS of (2.23) is obviously similar to A (they have the same eigenvalues). In the multi-dimensional case, with real and complex conjugate eigenvalues and eigenvectors (but no multiple eigenvalues), the matrix of coefficients A can be transformed into a similar block diagonal matrix with the general form

$$\begin{pmatrix} \ddots & \cdots & & & 0 \\ & \lambda_i & & & \\ \vdots & & \ddots & & \vdots \\ & & & B_j & \\ 0 & & \cdots & & \ddots \end{pmatrix}$$

where λ_i are real eigenvalues and B_j are (2×2) blocks of the same form as (2.23).

We conclude this section with a brief comment on the complications arising from repeated or multiple eigenvalues. Consider a simple system of two differential equations for which there is a double real eigenvalue $\lambda_1 = \lambda_2 = \lambda$, $A \neq \lambda I$, and therefore, the state space is not spanned by two linearly independent eigenvectors.[3] In this case, if u is an eigenvector of A, there is one solution with the usual form

$$x_1(t) = e^{\lambda t} u.$$

It can be verified that there is a second solution with the form

$$x_2(t) = e^{\lambda t}(tu + v). \tag{2.24}$$

The condition for (2.24) to be a solution is

$$\dot{x}_2(t) = e^{\lambda t}\left[u + \lambda(tu + v)\right] = Ax_2(t) = e^{\lambda t} A(tu + v)$$

or, because $e^{\lambda t} \neq 0$,

$$u = (A - \lambda I)(tu + v). \tag{2.25}$$

---

[3]The special case in which the (2×2) matrix $A = \lambda I$, λ real, sometimes called a bicritical node, has solutions that are half-lines from the origin, see figure 2.2(a).

Pre-multiplying by $(A - \lambda I)$ we have

$$(A - \lambda I)u = (A - \lambda I)^2(tu + v). \tag{2.26}$$

But in view of the Cayley–Hamilton theorem mentioned above and given that, in this case, $\text{tr}(A) = 2\lambda$ and $\det(A) = \lambda^2$, we have

$$A^2 - 2\lambda A + \lambda^2 = (A - \lambda I)^2 = 0.$$

Hence, (2.25) is satisfied if u is an eigenvector of A and v solves the equation $u = (A - \lambda I)v$. Then, by the superposition principle, the general solution can be written as

$$x(t) = c_1 x_1(t) + c_2 x_2(t) \tag{2.27}$$

and

$$x(0) = c_1 u + c_2 v$$

where c_1 and c_2 are two arbitrary constants depending on the initial conditions. Because v and u are linearly independent, they span the whole two-dimensional space. Thus, any arbitrary initial condition can be expressed as a linear combination of these two vectors. If the double eigenvalue $\lambda < 0$, the solution $x(t) \to 0$ as $t \to +\infty$ because the $\lim_{t \to \infty} e^{\lambda t}t = 0$, that is, the exponential contraction dominates the multiplicative effect of t. In the special case that the double eigenvalue $\lambda = 0$, the solution takes the form $x(t) = (c_1 u + c_2 v) + c_2 tu$, leading to an unstable equilibrium, divergence from which, however, grows linearly, not exponentially.

Solutions have more complicated forms if multiplicities are greater than two. Broadly speaking, when λ is a triple eigenvalue, the solution may include a term t^2, when it is a quadruple eigenvalue, a term t^3 and so on. The general formula for repeated eigenvalues is the following:

$$x(t) = \sum_{j=1}^{h} \sum_{l=0}^{n_j - 1} k_{jl} t^l e^{\lambda_j t} \tag{2.28}$$

where $n_j \geq 1$ is the multiplicity of the jth eigenvalue (so $n_j = 2$ means that $\lambda_j = \lambda_{j+1}$); $h \leq m$ is the number of distinct eigenvalues; $\sum n_j = m$, $(j = 1, \ldots, h)$. The vector-valued constants k_{jl} depend on the m arbitrary initial conditions and the independent vectors. For further details see Blanchard et al. (1998), pp. 282 ff.

The transformation of A also takes a slightly different form, called a (real) Jordan canonical form, when there are multiplicities and the eigenvectors do not span the entire space. The resulting matrix has diagonal blocks such as

$$\begin{pmatrix} \lambda & 1 \\ 0 & \lambda \end{pmatrix}$$

for a real, double eigenvalue λ, or

$$\begin{pmatrix} B_j & I \\ 0 & B_j \end{pmatrix}$$

where B_j is as before, I is a (2×2) identity matrix and 0 is a (2×2) null matrix, for a double complex conjugate pair. For higher-order multi-plicities of eigenvalues the Jordan form may have more complicated diago-nal blocks. For a detailed discussion of the Jordan form see, for example, Hirsch and Smale (1974).

2.3 Continuous systems in the plane

In applications, we very often encounter linear systems of two differential equations (or a differential equation of second order), either because the underlying model is linear or because it is linearised around an equilibrium point. Systems in two dimensions are particularly easy to discuss in full detail and give rise to a number of interesting basic dynamic configurations. Moreover, in practice, it is very difficult or impossible to determine *exact* values of the eigenvalues and eigenvectors for matrices of order greater than two. We therefore dedicate this section to continuous systems in the plane.

The general form can be written as

$$\begin{pmatrix} \dot{x} \\ \dot{y} \end{pmatrix} = A \begin{pmatrix} x \\ y \end{pmatrix} = \begin{pmatrix} a_{11} & a_{12} \\ a_{21} & a_{22} \end{pmatrix} \begin{pmatrix} x \\ y \end{pmatrix} \tag{2.29}$$

with $x, y \in \mathbb{R}$, a_{ij} real constants. If $\det(A) \neq 0$, the unique equilibrium, for which $\dot{x} = \dot{y} = 0$, is $x = y = 0$.

The characteristic equation is

$$0 = \det \begin{pmatrix} a_{11} - \lambda & a_{12} \\ a_{21} & a_{22} - \lambda \end{pmatrix}$$
$$= \lambda^2 - (a_{11} + a_{22})\,\lambda + (a_{11}a_{22} - a_{12}a_{21})$$
$$= \lambda^2 - \operatorname{tr}(A)\,\lambda + \det(A).$$

and the eigenvalues are

$$\lambda_{1,2} = \frac{1}{2}\left(\operatorname{tr}(A) \pm \sqrt{\Delta}\right) \tag{2.30}$$

where $\Delta \equiv ([\text{tr}(A)]^2 - 4\det(A))$ is called the **discriminant**. For system (2.29) the discriminant is

$$\Delta = (a_{11} + a_{22})^2 - 4(a_{11}a_{22} - a_{12}a_{21})$$
$$= (a_{11} - a_{22})^2 + 4a_{12}a_{21}.$$

The different types of dynamic behaviour of (2.29) can be described in terms of the two eigenvalues of the matrix A, which in planar systems can be completely characterised by the trace and determinant of A. In the following we consider nondegenerate equilibria (for which λ_1 and λ_2 are both nonzero), unless otherwise stated. We distinguish behaviours according to the sign of the discriminant.

CASE 1 $\Delta > 0$ Eigenvalues and eigenvectors are real. Solutions have the form

$$x(t) = c_1 e^{\lambda_1 t} u_1^{(1)} + c_2 e^{\lambda_2 t} u_2^{(1)}$$
$$y(t) = c_1 e^{\lambda_1 t} u_1^{(2)} + c_2 e^{\lambda_2 t} u_2^{(2)}.$$

Off-equilibrium, the slope of the orbits in the (x, y) plane asymptotically tend to a common, definite limit equal to

$$\lim_{t \to \infty} \frac{dy(t)}{dx(t)} = \lim_{t \to \infty} \frac{\dot{y}}{\dot{x}} = \lim_{t \to \infty} \frac{c_1 \lambda_1 e^{\lambda_1 t} u_1^{(2)} + c_2 \lambda_2 e^{\lambda_2 t} u_2^{(2)}}{c_1 \lambda_1 e^{\lambda_1 t} u_1^{(1)} + c_2 \lambda_2 e^{\lambda_2 t} u_2^{(1)}} \tag{2.31}$$

and either c_1 or c_2 is nonzero. Suppose that $\lambda_1 > \lambda_2$. Dividing throughout by $e^{\lambda_1 t}$ and taking the limit, we have

$$\lim_{t \to \infty} \frac{dy}{dx} = \frac{u_1^{(2)}}{u_1^{(1)}}.$$

This is the slope of the eigenvector associated with the **dominant** eigenvalue, that is, the largest eigenvalue, λ_1. It defines the direction with which orbits asymptotically converge to the fixed point for $t \to +\infty$ or $t \to -\infty$.

We have three basic subcases corresponding to figure 2.1(a), (b), (e), respectively (eigenvalues are plotted in the complex plane).[4]

(i) $\text{tr}(A) < 0$, $\det(A) > 0$. In this case, eigenvalues and eigenvectors are real and both eigenvalues are negative (say, $0 > \lambda_1 > \lambda_2$). The two-dimensional state space coincides with the stable eigenspace. The

---

[4]A special case is $\det(A) = 0$ for which $\lambda_1 = 0$, $\lambda_2 = \text{tr}(A)$. The solution corresponding to the zero eigenvalue is a constant ($c_1 e^{\lambda_1 t} = c_1$) and the dynamics are determined by the sign of $\text{tr}(A)$ as for cases (i) or (ii). In this case, however, there is no longer a unique equilibrium, but a continuum of equilibria whose locus is the straight line through the origin defined by $y/x = -(a_{11}/a_{12}) = -(a_{21}/a_{22})$.

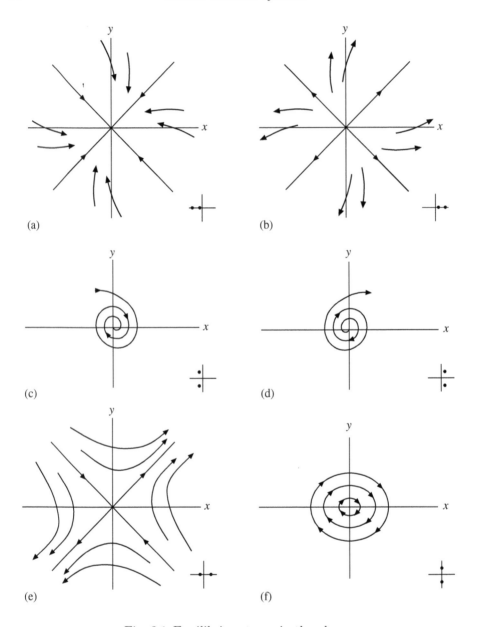

(a)

(b)

(c)

(d)

(e)

(f)

Fig. 2.1 Equilibrium types in the plane

equilibrium is called a **stable node**, the term 'node' referring to the characteristic shape of the ensemble of orbits around the equilibrium.

(ii) $\text{tr}(A) > 0$, $\det(A) > 0$. In this case, eigenvalues and eigenvectors are real, both eigenvalues are positive (say, $\lambda_1 > \lambda_2 > 0$) and the state space coincides with the unstable eigenspace. The equilibrium is called an **unstable node**.

(iii) $\det(A) < 0$. In this case, which implies $\Delta > 0$ independently of the sign of the trace of A, one eigenvalue is positive, the other is negative (say, $\lambda_1 > 0 > \lambda_2$). There is, then, a one-dimensional stable and a one-dimensional unstable eigenspace and the equilibrium is known as a **saddle point**. All orbits starting off-equilibrium eventually diverge from equilibrium except those originating in points on the stable eigenspace which converge to equilibrium.

remark 2.4 Mathematically speaking, a saddle point is unstable. In economic literature, especially in rational expectations or optimal growth models, we often find the concept of **saddle point stability** which, in the light of what we have just said, sounds like an oxymoron. The apparent contradiction is explained by considering that those models typically comprise a dynamical system characterised by a saddle point, *plus* some additional constraints on the dynamics of the system such that, given the initial condition of some of the variables, the others must be chosen so as to position the system on the stable set. Therefore, the ensuing dynamics is convergence to equilibrium.[5]

When eigenvectors are real, as in the cases of nodes and saddle points, the eigenvalues and eigenvectors can be given interesting interpretations. To see this, let us assume that the first element of each eigenvector is nonzero and let us fix the arbitrary scale factor so that, for each vector, the first element is equal to one. Therefore we can write

$$u_1 = (1, u_1^{(2)})^T$$
$$u_2 = (1, u_2^{(2)})^T.$$

In the two-dimensional case, the expansion of (2.10) gives

$$\begin{pmatrix} a_{11} & a_{12} \\ a_{21} & a_{22} \end{pmatrix} \begin{pmatrix} 1 \\ u_i^{(2)} \end{pmatrix} = \lambda_i \begin{pmatrix} 1 \\ u_i^{(2)} \end{pmatrix}$$

whence

$$a_{11} + a_{12} u_i^{(2)} = \lambda_i$$
$$a_{21} + a_{22} u_i^{(2)} = \lambda_i u_i^{(2)}$$

[5]See for example, Magill (1977), p. 190; Brock and Malliaris (1989), p. 125; Turnovsky (1996), p. 137.

and the second element of the eigenvector is easily calculated as

$$u_i^{(2)} = \frac{\lambda_i - a_{11}}{a_{12}} = \frac{a_{21}}{\lambda_i - a_{22}} \qquad i = 1, 2. \qquad (2.32)$$

Applying our discussion of section 2.2 to the simple two-dimensional case we can conclude that the quantities $u_i^{(2)}$ are equal to the slopes of the straight lines emanating from the origin in the plane (x, y) which are invariant under the law of motion defined by (2.29). From (2.29) the proportional rates of change of the two variables x, y are

$$\begin{aligned} \frac{d\ln x}{dt} &= \frac{\dot{x}}{x} = \frac{a_{11}x + a_{12}y}{x} = a_{11} + a_{12}\frac{y}{x} \\ \frac{d\ln y}{dt} &= \frac{\dot{y}}{y} = \frac{a_{21}x + a_{22}y}{y} = a_{21}\frac{x}{y} + a_{22}. \end{aligned} \qquad (2.33)$$

Using (2.33) we can see that for initial conditions on one or the other of the lines defined by (2.32), we have

$$\frac{\dot{x}}{x} = \frac{\dot{y}}{y} = \lambda_i$$

that is, the proportional rates of change for both variables are constant and equal to the corresponding eigenvalue. The lines defined by (2.32) are sometimes denoted as **balanced growth paths**. On these paths the two variables evolve in time maintaining a constant ratio. These paths can be of special significance in applications. For example, in an economic context, if y denotes capital stock and x is output, along a balanced growth path we have a growing (or declining) output with a constant capital/output ratio. Alternatively, if $y = \dot{x}$, and x is income, along the balanced growth path we have a constant proportional rate of growth (or decline) of income.

Notice that on the balanced growth paths the value of the variables may increase or decrease according to the sign of the eigenvalues *and* the initial conditions. A positive proportional growth rate may result either from positive rate of change ($\dot{x} > 0$) and positive level ($x > 0$) *or* from negative rate of change ($\dot{x} < 0$) and negative level ($x < 0$). A negative proportional growth rate results from the rate of change and level being of opposite sign. Finally, from the discussion above, it follows that the balanced growth path corresponding to the largest eigenvalue (in algebraic terms) is attracting in the sense that almost all orbits (i.e., all orbits except those starting on the other eigenvector) asymptotically converge to it in time.

CASE 2 $\Delta < 0$ The eigenvalues and eigenvectors are complex conjugate pairs and we have

$$(\lambda_1, \lambda_2) = (\lambda, \bar{\lambda}) = \alpha \pm i\beta$$

with

$$\alpha = \frac{1}{2}\,\mathrm{tr}(A) \qquad \beta = \frac{1}{2}\sqrt{-\Delta}.$$

As we have seen in the previous section, the solutions have the form

$$x(t) = Ce^{\alpha t}\cos(\beta t + \phi)$$
$$y(t) = Ce^{\alpha t}\sin(\beta t + \phi)$$

and the motion is oscillatory. If $\alpha \neq 0$ there is no strict periodicity in the sense that there exists no τ such that $x(t) = x(t + \tau)$. However, a **conditional period** can be defined as the length of time between two successive maxima of a variable, which is equal to $2\pi/\beta$. The **frequency** is simply the number of oscillations per time unit, that is, $\beta/2\pi$. The **amplitude** or size of the oscillations, depends on the initial conditions and $e^{\alpha t}$ (more on this point below).

The complex conjugate eigenvectors can be calculated as follows. We once again use the degree of freedom and, assuming that the real (denoted Re) and imaginary (denoted Im) parts are nonzero, set $u_1^{(1)} = 1+i$, $u_2^{(1)} = 1-i$. Then, from (2.10), we obtain

$$u_1 = \left(1+i, \frac{\alpha - \beta - a_{11}}{a_{12}} + i\frac{\alpha + \beta - a_{11}}{a_{12}}\right)^T$$
$$u_2 = \left(1-i, \frac{\alpha - \beta - a_{11}}{a_{12}} - i\frac{\alpha + \beta - a_{11}}{a_{12}}\right)^T. \tag{2.34}$$

From (2.34) we can construct two real and linearly independent vectors

$$e_R = \frac{1}{2}(u_1 + u_2) = \left(1, \frac{\alpha - \beta - a_{11}}{a_{12}}\right)^T$$
$$e_I = \frac{i}{2}(u_1 - u_2) = \left(-1, \frac{-\alpha - \beta + a_{11}}{a_{12}}\right)^T$$

which span a plane that is invariant. Notice that in this particular case, the invariant plane coincides with the state space and with the stable or unstable eigenspace. This is no longer the case in higher-dimensional systems (see, for example, figure 2.4(b)). There are three subcases depending on the sign of $\mathrm{tr}(A)$ and therefore of $\mathrm{Re}(\lambda) = \alpha$, see the corresponding figures in figure 2.1(c), (d), (f), respectively:

(i) $\operatorname{tr}(A) < 0$, $\operatorname{Re} \lambda = \alpha < 0$. The oscillations are dampened and the system converges to equilibrium. The equilibrium point is known as a **focus** or, sometimes, a **vortex**, due to the characteristic shape of the orbits around the equilibrium. In this case the focus or vortex is stable and the stable eigenspace coincides with the state space.

(ii) $\operatorname{tr}(A) > 0$, $\operatorname{Re} \lambda = \alpha > 0$. The amplitude of the oscillations gets larger with time and the system diverges from equilibrium. The unstable eigenspace coincides with the state space and the equilibrium point is called an unstable focus or vortex.

(iii) $\operatorname{tr}(A) = 0$, $\operatorname{Re} \lambda = \alpha = 0$, $\det(A) > 0$. In this special case we have a pair of purely imaginary eigenvalues. Orbits neither converge to, nor diverge from, the equilibrium point, but they oscillate regularly around it with a constant amplitude that depends only on initial conditions and a frequency equal to $\sqrt{\det(A)}/2\pi$. The centre eigenspace coincides with the state space and the equilibrium point is called a **centre**. We shall have more to say on this special case when we discuss nonlinear systems in chapters 3 and 5.

CASE 3 $\Delta = 0$ The eigenvalues are real and equal, $\lambda_1 = \lambda_2 = \lambda$. In this case, if $A \neq \lambda I$, only one eigenvector can be determined, call it $u = (u^{(1)}, u^{(2)})^T$, defining a single straight line through the origin. Recalling the form of the second solution from (2.24), we can write the general solution as

$$x(t) = (c_1 u^{(1)} + c_2 v^{(1)})e^{\lambda t} + t c_2 u^{(1)} e^{\lambda t}$$
$$y(t) = (c_1 u^{(2)} + c_2 v^{(2)})e^{\lambda t} + t c_2 u^{(2)} e^{\lambda t}$$

whence

$$x(0) = c_1 u^{(1)} + c_2 v^{(1)}$$
$$y(0) = c_1 u^{(2)} + c_2 v^{(2)}.$$

The equilibrium type is again a node, sometimes called a Jordan node. An example of this type is provided in figure 2.2(b), where it is obvious that there is a single eigenvector. If we choose $[x(0), y(0)]$ such that $c_1 \neq 0$, $c_2 = 0$, we have, for every $t \geq 0$

$$\frac{y(t)}{x(t)} = \frac{y(0)}{x(0)} = \frac{u^{(2)}}{u^{(1)}}$$

thus the straight line defined by u is invariant. Along that line, the system asymptotically converges to the equilibrium ($\lambda < 0$) or diverges from it ($\lambda > 0$). However, the vector v does not define a second invariant straight

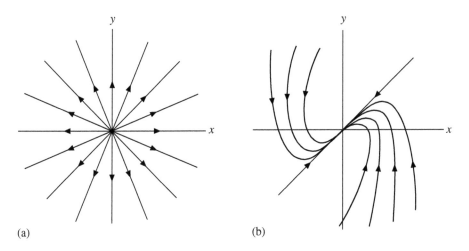

Fig. 2.2 Equilibrium types in the plane with repeated eigenvalue: (a) bicritical node; (b) Jordan node

line. To see this, choose the initial conditions so that $c_1 = 0$, $c_2 \neq 0$. Then at $t = 0$ we have

$$\frac{y(0)}{x(0)} = \frac{v^{(2)}}{v^{(1)}}$$

and at time $t \neq 0$

$$\frac{y(t)}{x(t)} = \frac{y(0)}{x(0)} + t\frac{u^{(2)}}{u^{(1)}}$$

which clearly is not time-invariant.

Finally, if $A = \lambda I$ the equilibrium is still a node, sometimes called a bicritical node. However, all half-lines from the origin are solutions, giving a star shaped form (see figure 2.2(a)).

The set of solution curves of a dynamical system, sketched in the state space, is known as the **phase diagram** or **phase portrait** of the system. Phase diagrams in two dimensions usually provide all the information we need on the orbit structure of the continuous-time system. The appendix explains the procedure to construct the phase diagram for typical cases of distinct eigenvalues and we refer the interested reader to it.

Finally, figure 2.3 provides a very useful geometric representation in the $(\mathrm{tr}(A), \det(A))$ plane of the various cases discussed above. Quadrants III and IV of the plane correspond to saddle points, quadrant II to stable nodes and foci and quadrant I to unstable nodes and foci. The parabola divides

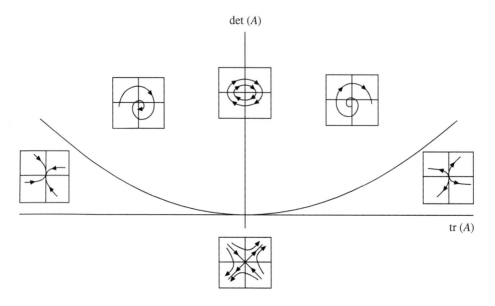

Fig. 2.3 Continuous-time dynamics in $\mathbb{R}^2$

quadrants I and II into nodes and foci (the former below the parabola, the latter above it). The positive part of the $\det(A)$-axis corresponds to centres.

The analysis for systems with $n > 2$ variables can be developed along the same lines although geometric insight will fail when the dimension of the state space is larger than three. In order to give the reader a broad idea of common situations we depict sample orbits of three-dimensional systems in figure 2.4, indicating the corresponding eigenvalues in the complex plane. The system in figure 2.4(a) has two positive real eigenvalues and one negative eigenvalue. The equilibrium point is an unstable saddle. All orbits eventually converge to the unstable eigenspace, in this case the plane associated with the positive real eigenvalues, and are captured by the expanding dynamics. The only exceptions are those orbits initiating on the stable eigenspace (defined by the eigenvector associated with the negative eigenvalue) which converge to the plane at the equilibrium point. Notice that the term *saddle* in $\mathbb{R}^m$ refers to all the cases in which there exist some eigenvalues with positive and some with negative real parts. We use the term **saddle node** when eigenvalues are all real, **saddle focus** when some of the eigenvalues are complex. An example of the latter is represented in figure 2.4(b). The two real vectors associated with the complex conjugate pair of eigenvalues, with negative real part, span the stable eigenspace. Orbits approach asymptotically the unstable eigenspace defined by the eigenvector associated with the positive eigenvalue, along which the dynamics is explosive.

2.4 General solutions of linear systems in discrete time

Let us consider now the linear, discrete-time system (2.5)

$$x_{n+1} = Bx_n \qquad x \in \mathbb{R}^m.$$

Once again we observe that $x = 0$ is an equilibrium solution and that for linear systems like (2.5) it is, generically, the only equilibrium. The general solution of (2.5) can be obtained following an argument entirely analogous to that used in section 2.2. If κ_i is a real, distinct eigenvalue of the $(m \times m)$ matrix B and v_i is the corresponding real eigenvector so that $Bv_i = \kappa_i v_i$, it can be verified that

$$x_i(n) = \kappa_i^n v_i \tag{2.35}$$

is a solution of (2.5). At this point we ought to mention an important difference between solutions of continuous-time systems and solutions of discrete-time systems regarding the status of the invariant lines defined by the eigenvectors. The real eigenvectors associated with real eigenvalues for systems of difference equations are still invariant sets, that is, sequences of points initiating on these sets remain on the sets. However the invariant, continuous, straight lines are *not solution orbits* as they were in the continuous case. Also recall from our discussion in chapter 1, section 1.2 the interesting peculiarity of discrete-time systems in the case of negative, real eigenvalues known as improper oscillations. In such a case, we may have the following interesting situation. Consider again system (2.5). Suppose we have one real eigenvalue $\kappa_1 < 0$, $|\kappa_1| > 1$ and that the initial condition is $x(0) = v_1$ (the eigenvector corresponding to κ_1). The solution is

$$x(n) = \kappa_1^n v_1 \tag{2.36}$$

that is, the system stays forever on the straight line through the origin defined by v_1. At each iterate the system moves towards the equilibrium point $\bar{x} = 0$, but always overshoots it, so that the distance from equilibrium becomes ever larger in time, tending to ∞ as $n \to +\infty$. Therefore, when there are negative real eigenvalues in discrete-time systems, great care should be applied in the use and, in particular, the interpretation of arrows in the phase diagram.

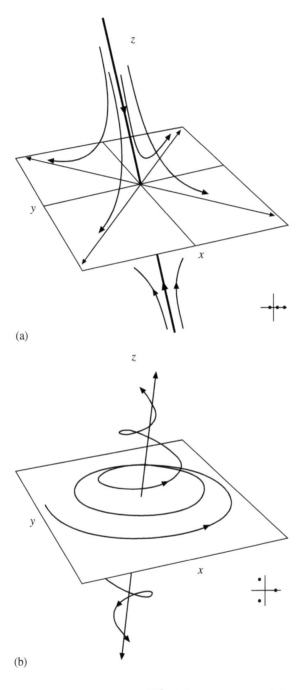

(a)

(b)

Fig. 2.4 Continuous-time systems in $\mathbb{R}^3$: (a) saddle node; (b) saddle focus

remark 2.5 Notice that if system (2.5) is characterised by improper oscillations then the system

$$x_{n+1} = B^2 x_n$$

with $B^2 = BB$, will have the same stability properties, but no improper oscillations. In fact, if κ is an eigenvalue of B, then $\rho = \kappa^2$ is an eigenvalue of B^2 and $|\rho| = |\kappa|^2$. Then $|\rho| \gtrless 1$ if $|\kappa| \gtrless 1$.

remark 2.6 If the (nonsingular) matrix B has an odd number of negative eigenvalues (and therefore $\det(B) < 0$), system (2.5) is said to be **orientation-reversing** (whereas for $\det(B) > 0$ it will be said to be **orientation-preserving**). The justification for these phrases can be readily seen in the simplest case of a one-dimensional system $x_{n+1} = G(x_n) = -x_n$, $x_n \in \mathbb{R}$. The oriented interval $[a, b]$ is mapped by G to an interval $[G(a), G(b)]$ with reversed orientation. However, $G^2[a, b]$ preserves orientation.

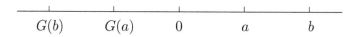

$$\begin{array}{ccccc} G(b) & G(a) & 0 & a & b \end{array}$$

As usual, dealing with complex conjugate eigenvalues is more complicated. Suppose that we have a pair of eigenvalues of B

$$(\kappa_j, \kappa_{j+1}) = (\kappa_j, \bar{\kappa}_j) = \sigma_j \pm i\theta_j$$

with a corresponding pair of eigenvectors

$$(v_j, v_{j+1}) = (v_j, \bar{v}_j) = p_j \pm iq_j$$

where p_j and q_j are m-dimensional vectors. Then the pair of functions

$$\begin{aligned} x_j(n) &= \frac{1}{2}\left(\kappa_j^n v_j + \bar{\kappa}_j^n \bar{v}_j\right) \\ &= r_j^n \left[p_j \cos(\omega_j n) - q_j \sin(\omega_j n)\right] \\ x_{j+1}(n) &= -\frac{i}{2}\left(\kappa_j^n v_j - \bar{\kappa}_j^n \bar{v}_j\right) \\ &= r_j^n \left[p_j \sin(\omega_j n) + q_j \cos(\omega_j n)\right], \end{aligned} \tag{2.37}$$

are solutions of (2.5). Here we have used the polar coordinate transformations:

$$\begin{aligned} \sigma_j &= r_j \cos \omega_j \\ \theta_j &= r_j \sin \omega_j \end{aligned} \tag{2.38}$$

and a well-known theorem due to De Moivre which states that

$$(\cos \omega \pm i \sin \omega)^n = \cos(\omega n) \pm i \sin(\omega n).$$

From (2.38) we have $r = \sqrt{\sigma^2 + \theta^2}$. Then r is simply the modulus of the complex eigenvalues. The solutions in (2.37) can be simplified in a manner analogous to that used for continuous-time systems. If $p_j^{(l)}$ and $q_j^{(l)}$ denote the lth elements of p_j and q_j, respectively, then in polar coordinates we have

$$p_j^{(l)} = C_j^{(l)} \cos(\phi_j^{(l)})$$
$$q_j^{(l)} = C_j^{(l)} \sin(\phi_j^{(l)})$$

where $l = 1, 2, \ldots, m$. Using again the trigonometric identities in (2.15), the solutions (2.37) can be re-written as m-dimensional vectors whose lth elements have the form

$$x_j^{(l)}(n) = C_j^{(l)} r_j^n \cos(\omega_j n + \phi_j^{(l)})$$
$$x_{j+1}^{(l)}(n) = C_j^{(l)} r_j^n \sin(\omega_j n + \phi_j^{(l)}).$$

Notice the similarities and differences with solutions to continuous-time systems with complex eigenvalues in (2.16).

Stable, unstable and centre eigenspaces are defined in a manner similar to that used for continuous-time systems, with the obvious difference that the discriminating factor for discrete-time systems is whether the *moduli* of the relevant eigenvalues are, respectively, smaller, greater or equal to *one*.

Once again, assuming that we have m linearly independent solutions defined by (2.35) and (2.37), by the superposition principle the general solution of (2.5) can be written as a linear combination of the individual solutions as in (2.17), namely:

$$x(n) = c_1 x_1(n) + c_2 x_2(n) + \cdots + c_m x_m(n) \tag{2.39}$$

where c_i are constants depending on the initial conditions.

As in the continuous case, a complication arises when eigenvalues are repeated. The general solution analogous to (2.28) is

$$x(n) = \sum_{j=1}^{h} \sum_{l=0}^{n_j - 1} k_{jl} n^l \kappa_j^n$$

where $n_j \geq 1$ is the multiplicity of the jth eigenvalue, $h \leq m$ is the number of distinct eigenvalues.

Inspection of (2.35) and (2.37) indicates that if the modulus of *any* of the eigenvalues is greater than one, solutions tend to $+$ or $-\infty$ as time tends to plus infinity. On the contrary, if *all* eigenvalues have modulus smaller than one, solutions converge asymptotically to the equilibrium point. In the latter case, we say that the equilibrium is asymptotically stable, in the former case we say it is unstable. For distinct eigenvalues, the solution in the long run is determined by the term (or terms in the complex case) corresponding to the eigenvalue (or the pair of eigenvalues) with the largest modulus. When the dominant real eigenvalue or complex eigenvalue pair has modulus exactly equal to one, we have a weaker form of stability which we discuss in chapter 3. If that eigenvalue or eigenvalue pair is repeated, the equilibrium is unstable. As for continuous-time systems, there are criteria that can be applied to the coefficient matrix B to ensure stability. In this case, stability conditions guarantee that all eigenvalues have moduli inside the unit circle of the complex plane. There is also a compact form for the general solution of (2.5). We proceed as we did for the continuous-time case. Let X_n be the matrix whose columns are the m solutions defined by (2.35) and (2.37). We can verify that

$$X_n = B^n X_0$$

where $X_n = X(n)$ and $X_0 = X(0)$. Then

$$x_n = X_n c = B^n x_0$$

where $c = (c_1, c_2, \ldots, c_m)^T$ are the constants of (2.39) and $x_n = x(n)$, $x_0 = x(0)$. To verify this consider that

$$x_{n+1} = B^{n+1} x_0 = BB^n x_0 = Bx_n$$

that is, we have (2.5).

As in the case of continuous-time systems, we often encounter discrete-time linear systems in two variables, either because the original system is linear or because we are studying a linearised system in two dimensions. We now turn to a detailed discussion of these.

2.5 Discrete systems in the plane

The discrete autonomous system analogous to continuous system (2.29) is:

$$\begin{pmatrix} x_{n+1} \\ y_{n+1} \end{pmatrix} = B \begin{pmatrix} x_n \\ y_n \end{pmatrix} = \begin{pmatrix} b_{11} & b_{12} \\ b_{21} & b_{22} \end{pmatrix} \begin{pmatrix} x_n \\ y_n \end{pmatrix}. \tag{2.40}$$

If the matrix $(I - B)$ is nonsingular, which we assume here, there exists a unique equilibrium point for (2.40), situated at the origin. The characteristic equation is analogous to the continuous case, that is,

$$\kappa^2 - \text{tr}(B)\,\kappa + \det(B) = 0$$

and the eigenvalues are

$$\kappa_{1,2} = \frac{1}{2}\left(\text{tr}(B) \pm \sqrt{[\text{tr}(B)]^2 - 4\det(B)}\right).$$

The dynamics for the discrete system (2.40) are discussed case by case, again considering nondegenerate equilibria ($|\kappa_1|, |\kappa_2| \neq 1$):

CASE 1 $\Delta > 0$ The eigenvalues are real and solutions take the form

$$x(n) = c_1 \kappa_1^n v_1^{(1)} + c_2 \kappa_2^n v_2^{(1)}$$
$$y(n) = c_1 \kappa_1^n v_1^{(2)} + c_2 \kappa_2^n v_2^{(2)}.$$

(i) If $|\kappa_1| < 1$ and $|\kappa_2| < 1$ the fixed point is a stable node. In discrete time this means that solutions are sequences of points approaching the equilibrium as $n \to +\infty$. If $\kappa_1, \kappa_2 > 0$ the approach is monotonic, otherwise, there are improper oscillations (see figures 2.5(a) and (c), respectively). In this case, the stable eigenspace coincides with the state space.

(ii) If $|\kappa_1| > 1$ and $|\kappa_2| > 1$ the fixed point is an unstable node, that is, solutions are sequences of points approaching equilibrium as $n \to -\infty$. If $\kappa_1, \kappa_2 > 0$ the approach is monotonic, otherwise, there are improper oscillations (as in figures 2.5(a) and (c), respectively, but arrows point in the opposite direction and the time order of points is reversed). In this case, the unstable eigenspace coincides with the state space.

(iii) If $|\kappa_1| > 1$ and $|\kappa_2| < 1$ the fixed point is a saddle point. No sequences of points approach the equilibrium for $n \to \pm\infty$ except those originating from points on the eigenvectors associated with κ_1 or κ_2. Again, if $\kappa_1, \kappa_2 > 0$ orbits move monotonically (see figure 2.5(b)), otherwise they oscillate improperly (see figure 2.5(d)). The stable and unstable eigenspaces are one-dimensional.

CASE 2 $\Delta < 0$ Then $\det(B) > 0$ and eigenvalues are a complex conjugate pair

$$(\kappa_1, \kappa_2) = (\kappa, \bar{\kappa}) = \sigma \pm i\theta$$

and solutions are sequences of points situated on spirals whose amplitude increases or decreases in time according to the factor r^n where $r = |\sigma \pm i\theta| = $

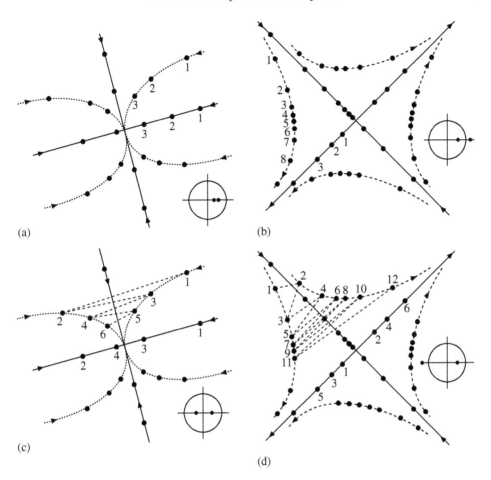

Fig. 2.5 Phase diagrams with real eigenvalues: (a) and (c) stable nodes; (b) and (d) saddle points

$\sqrt{\sigma^2 + \theta^2} = \sqrt{\det(B)}$, the modulus of the complex eigenvalue pair. The invariant plane in the two-dimensional problem coincides with the state space as well as the (stable, unstable or centre) eigenspace. Solutions are of the form

$$x(n) = Cr^n \cos(\omega n + \phi)$$
$$y(n) = Cr^n \sin(\omega n + \phi).$$

(i) If $r < 1$ solutions converge to equilibrium and the equilibrium point is a stable focus (see figure 2.6(a)).

(ii) If $r > 1$ solutions diverge and the equilibrium point is an unstable focus (as in figure 2.6(a), but arrows point in the opposite direction and the time order of points is reversed).

(iii) If $r = 1$ the eigenvalues lie exactly on the unit circle, an exceptional case. There are two subcases which depend on the frequency of the oscillation $\omega/2\pi$, $\omega = \arccos[\text{tr}(B)/2]$:

(a) $\omega/2\pi$ is rational and the orbit in the state space is a periodic sequence of points situated on a circle, the radius of which depends on initial conditions (see figure 2.6(b));

(b) $\omega/2\pi$ is irrational, the sequence is nonperiodic or **quasiperiodic**. A formal definition is delayed until chapter 4 but we can describe the sequence as follows. Starting from any point on the circle, orbits stay on the circle but no sequence returns to the initial point in finite time. Therefore, solutions *wander* on the circle *filling it up*, without ever becoming periodic (see figure 2.6(c)). We shall have more to say on quasiperiodic solutions, both in discrete and continuous time when we discuss nonlinear systems, for which periodic and quasiperiodic solutions are not exceptional.[6]

CASE 3 $\Delta = 0$ There is a repeated real eigenvalue $\kappa_1 = \text{tr}(B)/2$. Considerations analogous to those made for continuous systems could be developed to arrive at the general form of solutions to planar systems with a repeated eigenvalue κ, namely

$$x(n) = (c_1 v^{(1)} + c_2 u^{(1)})\kappa^n + n c_2 v^{(1)} \kappa^n$$
$$y(n) = (c_1 v^{(2)} + c_2 u^{(2)})\kappa^n + n c_2 v^{(2)} \kappa^n.$$

If $|\kappa| < 1$, $\lim_{n \to \infty} n\kappa^n = 0$ and the multiplicative expansion due to n is dominated by the exponential contraction implied by κ^n. If the repeated eigenvalue is equal to one in absolute value, the equilibrium is unstable (with improper oscillations for $\kappa_1 = -1$). However, divergence is linear not exponential.

The equilibrium types illustrated in figures 2.5 and 2.6 are presented in such a way as to emphasise the differences with equilibrium types for continuous planar systems. First, it should be observed that the points of the orbit sequences lie on the (dotted) curves but the curves themselves are not

---

[6]The possibility of quasiperiodic orbits also exists in linear continuous-time systems, in the exceptional case for which two or more pairs of complex eigenvalues have zero real parts. If two or more frequencies are incommensurable, the motion is quasiperiodic, otherwise it is periodic (cf. chapter 4, section 4.5).

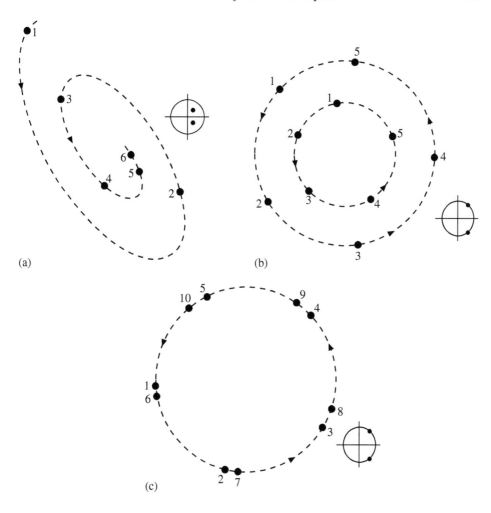

(a)

(b)

(c)

Fig. 2.6 Phase diagrams with complex eigenvalues: (a) a stable focus; (b) periodic
cycles; (c) a quasiperiodic solution

orbits. Moreover, points have been given a time reference so as to avoid
any confusion about the direction of motion. In figure 2.5(a) and (c) a stable node is represented, for which eigenvalues are less than one in absolute
value. In (a) both eigenvalues are positive. In (c) one eigenvalue is positive, one negative and improper oscillations occur as the orbit jumps from
one curve to the other. Eigenvectors are invariant sets, and orbits on that
associated with the positive eigenvalue tend monotonically to the equilibrium. However, on the eigenvector associated with the negative eigenvalue,
the orbit jumps from one side of the equilibrium to the other. Similarly, in
figure 2.5(b) and (d) a saddle point is represented. In (b) both eigenvalues

are positive. In (d), one eigenvalue is less than minus one and, again, orbits jump from curve to curve unless they begin on one of the eigenvectors. Notice, in particular, that the motion on the eigenvector representing the unstable direction is *towards* equilibrium, as if it were to converge. However the orbit overshoots the equilibrium, with ever greater distance, at each iteration.

In the two-dimensional case the conditions for stability, that is for which $|\kappa| < 1$, can be given a simple representation in terms of the trace and determinant of the constant matrix B as follows

(i) $$1 + \text{tr}(B) + \det(B) > 0$$
(ii) $$1 - \text{tr}(B) + \det(B) > 0$$
(iii) $$1 - \det(B) > 0.$$

The dynamics for the discrete case can be conveniently summarised by the diagram in figure 2.7. (For simplicity's sake we represent orbits as continuous rather than dotted curves.) If we replace the greater-than sign with the equal sign in conditions (i)–(iii), we obtain three lines intersecting in the $(\text{tr}(B), \det(B))$ plane, defining a triangle. Points inside the triangle correspond to stable combinations of the trace and determinant of B.[7] The parabola defined by

$$\text{tr}(B)^2 = 4\det(B)$$

divides the plane in two regions corresponding to real eigenvalues (below the parabola) and complex eigenvalues (above the parabola). Combinations of trace and determinant above the parabola but in the triangle lead to stable foci, combinations below the parabola but in the triangle are stable nodes. All other combinations lead to unstable equilibria.

Finally in figure 2.8 an example of a system in $\mathbb{R}^3$ is provided. There are a complex conjugate pair with modulus less than one, and one dominant real eigenvalue greater than one. The equilibrium point is a saddle focus.

2.6 An economic example

In chapter 1, section 1.2 we discussed a model of partial economic equilibrium with a price-adjustment mechanism described by the equations

---

[7]If $1 + \text{tr}(B) + \det(B) = 0$ while (ii) and (iii) hold, one eigenvalue is equal to -1; if $1 - \text{tr}(B) + \det(B) = 0$ while (i) and (iii) hold, one eigenvalue is equal to $+1$; and if $1 = \det(B)$ while (i) and (ii) hold, the two eigenvalues are a complex conjugate pair with modulus equal to 1.

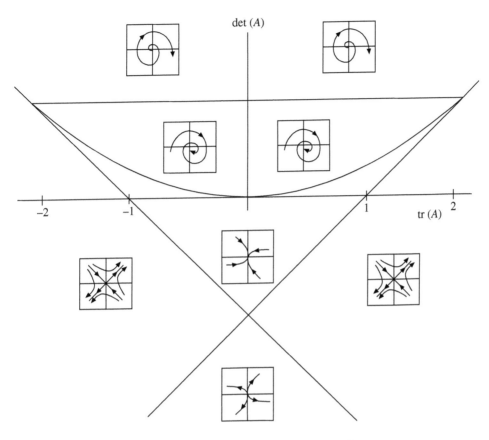

Fig. 2.7 Discrete-time dynamics in $\mathbb{R}^2$

$$D(p_n) = a - bp_n$$
$$S(p_n) = -\!-m + sp_n$$
$$p_{n+1} = p_n + [D(p_n) - S(p_n)]$$

with $a, b, m, s > 0$. In this section we study a modified version of the model.
Suppose that there is an asymmetric reaction to a price change on the part of the consumer and the producer. The former adjusts demand to price changes immediately, whereas the latter needs time to adjust supply. We represent this situation by a one-period discrete lag in the supply function. The modified supply function is then

$$S(p_{n-1}) = -m + sp_{n-1}. \tag{2.41}$$

After substituting the demand function and (2.41) into the price adjustment

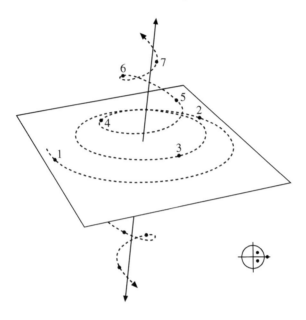

Fig. 2.8 A discrete-time system in $\mathbb{R}^3$

equation we have

$$p_{n+1} = p_n + (a - bp_n + m - sp_{n-1})$$
$$= (1 - b)p_n - sp_{n-1} + (a + m). \tag{2.42}$$

Equation (2.42) is a second-order difference equation. To transform it into a system of two first-order, homogeneous difference equations with the same dynamics we first reduce the order of the equation by introducing an auxiliary variable as the one-period lagged value of the original variable

$$z_n = p_{n-1}. \tag{2.43}$$

Next, to render the system homogeneous, we introduce an auxiliary variable which places the equilibrium at the origin of the coordinate axes,

$$\tilde{z}_n = \tilde{p}_{n-1} = p_{n-1} - \frac{a + m}{b + s}. \tag{2.44}$$

Substituting (2.43) and (2.44) into (2.42) we obtain the first-order system in two variables

$$\tilde{z}_{n+1} = \tilde{p}_n$$
$$\tilde{p}_{n+1} = -s\tilde{z}_n + (1 - b)\tilde{p}_n. \tag{2.45}$$

The only equilibrium point of (2.45) is $(0,0)$ from which we easily derive the equilibrium values of the original variable p, namely

$$\bar{p} = \frac{a + m}{b + s},$$

the same as the equilibria of the simpler cases discussed in chapter 1, section 1.2. Since the auxiliary variables $(\tilde{z}_n, \tilde{p}_n)$ are obtained by simple translation of (z_n, p_n), the dynamics of system (2.45) are the same as that of (2.42), only the equilibrium is translated by an amount equal to $\bar{p}$. The dynamics is then completely characterised by the eigenvalues of the constant matrix

$$B = \begin{pmatrix} 0 & 1 \\ -s & 1 - b \end{pmatrix} \tag{2.46}$$

which are the roots of the characteristic equation

$$\kappa^2 - (1 - b)\kappa + s = 0$$

namely,

$$\kappa_{1,2} = \frac{1}{2}\left[(1 - b) \pm \sqrt{(1 - b)^2 - 4s}\right].$$

Since there are, again, only two parameters in the characteristic polynomial, a full description of the dynamical behaviour can be graphically represented in the (b, s) plane, considering only the positive quadrant, see figure 2.9.

Consider first the type of motion, trigonometric oscillation or otherwise. The eigenvalues are real if $(1 - b)^2 > 4s$, complex if $(1 - b)^2 < 4s$. The curve that divides the parameter space is $s = (1 - b)^2/4$. The area inside the parabola represents combinations of parameters for which $s > (1-b)^2/4$ leading to oscillatory solutions (fixed points are foci or centres). For parameter combinations outside of the parabola there are monotonic dynamics or improper oscillations (fixed points are nodes or saddles).

The conditions for stability discussed in section 2.5 for this economic model are

(i) $\qquad\qquad\qquad\qquad\qquad 2 - b + s > 0$

(ii) $\qquad\qquad\qquad\qquad\qquad b + s > 0$

(iii) $\qquad\qquad\qquad\qquad\qquad 1 - s > 0.$

The second condition is always satisfied, given $s, b > 0$ by assumption. The first and third conditions must both be satisfied for stability. The area of stable behaviour is bounded by $s = 1$, $s = b - 2$, $s = 0$ and $b = 0$. Combinations of s and b within that area give equilibria that are either stable foci (above the parabola) or stable nodes (below the parabola).

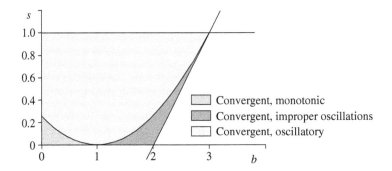

Fig. 2.9 Parameter space for the partial equilibrium model with a lag

If the discriminant is positive, $\det(B) = s > 0$ and $\operatorname{tr}(B) = 1-b$, there are either two positive real roots or two negative real roots, below the parabola, according to whether $b < 1$ or $b > 1$, respectively. Stable nodes with $b > 1$ are characterised by improper oscillations. If the discriminant is negative there are two complex conjugate roots with modulus $\sqrt{s}$. Trigonometric oscillations are dampened, persistent or explosive according to whether $\sqrt{s} \gtrless 1$.

How did the assumption of a lag in the suppliers' reaction to price changes alter the dynamics? The fact that (2.42) is second-order (or, equivalently, (2.45) is two-dimensional) introduces the possibility of *proper* oscillations. Moreover, the area of monotonic convergence is greatly reduced in the present model (cf. figure 1.3) and the stable dynamics is likely to be oscillatory. The combinations of parameter values leading to stability have changed. A simple lag in the response to price changes on the part of suppliers may modify the dynamics of the model significantly.

Appendix: phase diagrams

A phase diagram is a geometric representation of the orbit structure of a dynamical system and consequently a very useful tool of analysis of its qualitative properties. Given its geometric character, a phase diagram can be used only to investigate dynamical systems of one, two, or perhaps three dimensions. In this appendix we describe in some detail the procedure to construct phase diagrams of those linear continuous-time systems in $\mathbb{R}^2$ which have nonzero and nonrepeated eigenvalues. Many of the ideas and graphic methods can be extended to nonlinear systems in $\mathbb{R}^2$, although the exercise may become much more complicated.

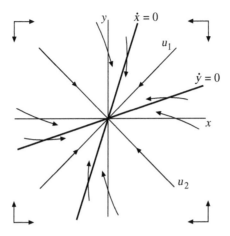

Fig. 2.10 Stable node: $a_{11}, a_{22} < 0$; $a_{12}, a_{21} > 0 \to \Delta > 0$; $a_{11}a_{22} > a_{12}a_{21}$

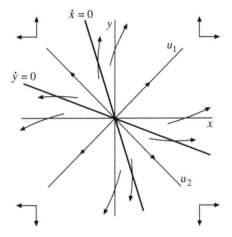

Fig. 2.11 Unstable node: $a_{11}, a_{22} > 0$; $a_{12}, a_{21} > 0 \to \Delta > 0$; $a_{11}a_{22} > a_{12}a_{21}$

The basic system under investigation has the general form

$$\begin{pmatrix} \dot{x} \\ \dot{y} \end{pmatrix} = A \begin{pmatrix} x \\ y \end{pmatrix} = \begin{pmatrix} a_{11} & a_{12} \\ a_{21} & a_{22} \end{pmatrix} \begin{pmatrix} x \\ y \end{pmatrix}$$

with the a_{ij} scalar constants and $x, y \in \mathbb{R}$. We consider the general case in which there is a unique equilibrium point. As we proceed through the steps the reader should refer to figures 2.10–2.14.

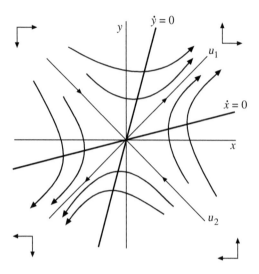

Fig. 2.12 Saddle point: $a_{11}, a_{22} < 0$; $a_{12}, a_{21} > 0 \to \Delta > 0$; $a_{11}a_{22} < a_{12}a_{21}$

STEP 1 Draw the two straight lines in the (x, y) plane, through the origin, defined by

$$\dot{x} = a_{11}x + a_{12}y = 0$$
$$\dot{y} = a_{21}x + a_{22}y = 0$$

which are, respectively

$$y = -\frac{a_{11}x}{a_{12}} \qquad (2.47)$$

$$y = -\frac{a_{21}x}{a_{22}}. \qquad (2.48)$$

The lines defined by (2.47) and (2.48) are called **null-clines** or **stationaries** as they are the sets of points in the (x, y) plane such that the velocity (rate of change with respect to time) of one or the other of the two variables is zero. In other words, if y is measured on the ordinate and x on the abscissa, at points on the null-clines, the tangent to an orbit is horizontal ($\dot{y} = 0$) or vertical ($\dot{x} = 0$). If A is nonsingular the null-clines intersect only at the origin, for which $\dot{x} = \dot{y} = 0$ and this is, by definition, the unique equilibrium of the system. The null-clines divide the (x, y) plane into four regions in which orbits have four corresponding different broad directions. In the (x, y) plane these directions may be imagined as 'northeast' (up and right); 'northwest' (up and left); 'southwest' (down and left); 'southeast' (down and right).

STEP 2 Establish the directions of orbits in each of the four regions by calculating the sign of the following partial derivatives:

$$\frac{\partial \dot{x}}{\partial x} = a_{11} \tag{2.49}$$

$$\frac{\partial \dot{y}}{\partial x} = a_{21} \tag{2.50}$$

or, alternatively and equivalently

$$\frac{\partial \dot{x}}{\partial y} = a_{12} \tag{2.51}$$

$$\frac{\partial \dot{y}}{\partial y} = a_{22}. \tag{2.52}$$

Equations (2.47)–(2.48) plus (2.49)–(2.50) (or (2.51)–(2.52)) give us all the necessary information to determine the basic slopes of orbits on any point of the plane. The direction of the flow in each area is usually indicated by pairs of arrows as in figures 2.10–2.14.

STEP 3 *If the eigenvalues are real*, draw the two straight lines corresponding to the two real eigenvectors. If $u_i^{(j)}$ is the jth element of the ith eigenvector, and we set $u_i^{(1)} = 1$ for $i = 1, 2$, then the slopes of the straight lines are equal to the second elements, which according to (2.32) are

$$u_i^{(2)} = \frac{\lambda_i - a_{11}}{a_{12}} = \frac{a_{21}}{\lambda_i - a_{22}} \qquad i = 1, 2.$$

(If the first element of an eigenvector is zero, then we set the second element to one.) In drawing the lines defined in (2.32) care should be taken to place them correctly in the plane, in relation to the null-clines. As already mentioned in the main text of this chapter, these lines are invariant in the sense that orbits starting on them remain on them forever.

At this point, the general picture of the orbit structure of the system is complete and often the diagram provides all of the information that we need. Phase diagrams of the canonical types of continuous-time systems in $\mathbb{R}^2$ are represented in figures 2.10–2.14 where the hypotheses on parameter signs are provided. The best way of learning the technique is for the reader to re-draw all of them for different sign patterns of the parameters.

remark 2.7 A similar procedure can also be used for a nonlinear system of differential equations $\dot{x} = f(x)$ $x \in \mathbb{R}^2$. However, keep in mind the following points:

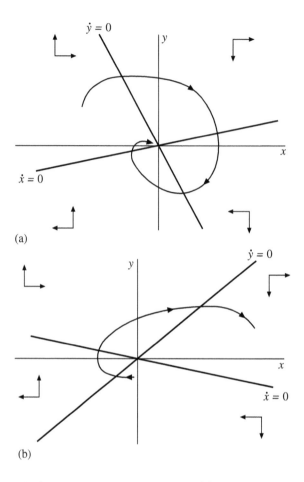

$\dot{y} = 0$

y

$\dot{x} = 0$

x

(a)

$\dot{y} = 0$

y

x

$\dot{x} = 0$

(b)

Fig. 2.13 Foci with $\Delta < 0$; $a_{11}a_{22} > 0 > a_{12}a_{21}$: (a) stable focus: $a_{11}, a_{21}, a_{22} < 0$ and $a_{12} > 0$; (b) unstable focus: $a_{11}, a_{12}, a_{22} > 0$, $a_{21} < 0$

(1) Null-clines defined by $f_i(x) = 0$ ($i = 1, 2$) are generally *not* straight lines. They may intersect an indefinite number of times. None of the intersections need be the origin of the coordinate axes. They also may not intersect at all (i.e., equilibria may not exist).

(2) Consequently, the state space may be subdivided into an indefinitely large number of regions, characterised by different slopes of the orbits.

(3) The derivatives (2.49)–(2.52) provide information on the directions of the orbits only locally.

(4) In general, there need not be any orbits which are straight lines. However, as we discuss in chapter 3, the linearised system around a hyperbolic equilibrium provides a good local approximation of the nonlinear

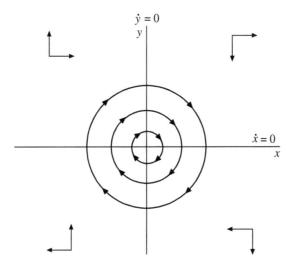

Fig. 2.14 Centre: $a_{11} = a_{22} = 0$; $a_{12} > 0$; $a_{21} < 0$; $\Delta < 0$; $a_{11}a_{22} > a_{12}a_{21}$

system, in a neighbourhood of equilibrium. Therefore, the phase diagram of that (linear) system can be used to represent the local behaviour of the true, nonlinear system. Sometimes different local diagrams can be connected so as to provide a useful approximation of the global behaviour.

(5) As we shall see later, in nonlinear systems in $\mathbb{R}^2$ there may exist special orbits that connect two equilibria, that move away from one equilibrium point and return to it, or that connect an equilibrium point with a periodic orbit. More complicated situations may arise in higher dimensions.

Exercises

2.1   Prove that the superposition principle is valid for (2.2) if, and only if, G is linear.

2.2   Verify that (2.13) are solutions of (2.4).

2.3   Prove that (2.14) are actually solutions of (2.4).

2.4   Verify property (i) of the fundamental matrix solution.

2.5   Verify (2.34).

2.6   Verify that (2.35) are solutions of (2.5).

2.7   Consider the autonomous system

$$\begin{pmatrix} \dot{x} \\ \dot{y} \end{pmatrix} = \begin{pmatrix} 1 & 2 \\ 3 & 0 \end{pmatrix} \begin{pmatrix} x \\ y \end{pmatrix}$$

(a) Find the fixed points, eigenvalues and associated eigenvectors.

(b) Plot the stationaries in the state space and indicate the motion in the regions defined by them.

(c) Sketch a phase diagram for the system.

2.8   The following system has a repeated eigenvalue. Write the exact solution using the procedure described in section 2.2 and the initial conditions $x(0) = 1$, $y(0) = 2$

$$\dot{x} = -x + y$$
$$\dot{y} = -x - 3y.$$

2.9   For the following systems of differential equations, determine the fixed point and its type

(a)

$$\dot{x} = 5x + 10$$
$$\dot{y} = 2y$$

(b)

$$\dot{x} = -\frac{1}{2}x - 2$$
$$\dot{y} = +\frac{1}{2}y - 2$$

(c)

$$\dot{x} = -0.8x$$
$$\dot{y} = -1.2y.$$

2.10   The following systems of difference equations are similar to the systems of differential equations in the previous problem. For each system, find the fixed point and determine its type. Compare the dynamics of each system with the corresponding one of exercise 2.9

(a)

$$x_{n+1} = 5x_n + 10$$
$$y_{n+1} = 2y_n$$

(b)

$$x_{n+1} = -\frac{1}{2}x_n - 2$$

$$y_{n+1} = +\frac{1}{2}y_n - 2$$

(c)

$$x_{n+1} = -0.8x_n$$

$$y_{n+1} = -1.2y_n.$$

2.11   Prove that if $(\kappa, \bar{\kappa}) = \sigma \pm i\theta$ is a complex eigenvalue pair with a corresponding pair of eigenvectors $(v, \bar{v}) = p \pm iq$ of the matrix B, then $\kappa^n v$ is a solution to (2.5).

2.12   Suppose that each of the following is the coefficient matrix B for a system of difference equations

$$\begin{pmatrix} 5 & -3 \\ -3 & 5 \end{pmatrix} \qquad \begin{pmatrix} 0 & -1 \\ 4 & 0 \end{pmatrix}$$

$$\begin{pmatrix} 0.2 & 0 & 0 \\ 0 & 0.5 & 0 \\ 0 & 0 & 1.5 \end{pmatrix} \qquad \begin{pmatrix} 1 & -1 & 0 \\ -1 & 2 & -1 \\ 0 & -1 & 1 \end{pmatrix}$$

(a) Find the eigenvalues of B and describe the corresponding dynamics.

(b) Find the associated eigenvectors for the first three matrices.

(c) Sketch the state space for the first three matrices.

2.13   This exercise was given as an example in a textbook where the equilibrium was classified as a *saddle point*

$$\begin{pmatrix} x_{n+1} \\ y_{n+1} \end{pmatrix} = \begin{pmatrix} 4 & 4 \\ 1 & -3 \end{pmatrix} \begin{pmatrix} x_n \\ y_n \end{pmatrix} + \begin{pmatrix} -8 \\ 4 \end{pmatrix}$$

(a) Perform a variable transformation so as to render the system homogeneous.

(b) Find the eigenvalues of the transformed system to two decimal places. Classify the equilibrium. Is this a saddle point?

(c) Determine the associated eigenvectors to two decimal places and plot them in the state space.

(d) Choose an initial value *on each eigenvector* and indicate an approximate sequence of points on that eigenvector. Why do you think the point was classified as a saddle point?

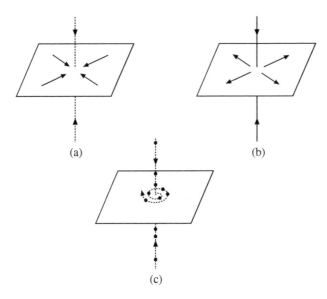

Fig. 2.15 Phase diagrams sketched in $\mathbb{R}^3$ for discrete- and continuous-time
systems

(e) Let the transformed variables have initial conditions of -1 and
1, respectively. Use these values, the eigenvalues and the eigenvectors to write down the exact solution to the transformed
system. Now calculate values for the first three iterations, for
the tenth and the hundredth iteration, to see how orbits move
through time.

2.14   Recalling remark 2.1 and adapting it to the discrete-time case, show
how we obtained the appropriate auxiliary variables (2.43) and (2.44)
for the second-order equation (2.42).

2.15   For the following exercises consider the autonomous system (2.45),
and the associated diagram of the parameter space in figure 2.9

(a) Choose two pairs of values of (b, s), holding b constant and varying s, in such a way that the equilibrium point of the resulting
model is (i) a stable node and (ii) an unstable node. In each
case determine the matrix B, the characteristic equation, the
eigenvalues and associated eigenvectors. Let $\tilde{z}_0 = 1$ $\tilde{p}_0 = 2$.
Write the exact solution for each parameter choice and calcu-
late values of $[\tilde{z}(n), \tilde{p}(n)]$ for the first three iterations, for the
tenth and the hundredth iteration.

(b) Choose values of (b, s) for which the equilibrium point is a stable focus. Calculate the complex eigenvalues and eigenvectors and the associated two real vectors.

2.16   Phase diagrams for three systems are sketched in figure 2.15. For each of them locate the eigenvalues in the complex plane. Roughly sketch some sample orbits. Give the dimensions of the stable and unstable eigenspaces.

3

Stability of fixed points

In chapter 2 we discussed stable and unstable equilibrium (or fixed) points of dynamical systems in continuous and discrete time by making use of a broad and intuitive notion of stability. In this chapter, we re-examine the issue and provide more precise and formal definitions. We also explain two fundamental methods for determining stability of equilibrium points of nonlinear systems. In chapter 4, our discussion will be extended to stability of sets 'larger' than a single point. Before proceeding, it is important to recall that nonlinear systems may be characterised by a plurality of fixed points (as well as other more complicated invariant sets). In those cases we cannot speak of stability, or instability, of the system as a whole, but stability properties of each point (or set) must be established separately. The global orbit structure of the system may sometimes be constructed by connecting the dynamically different regions of the state space. In the following sections, stability will be discussed for dynamical systems in $\mathbb{R}^m$, but most of the definitions and results could be extended to arbitrary metric spaces (spaces endowed with a distance function).

3.1 Some formal definitions of stability

Consider the usual autonomous system of differential equations

$$\dot{x} = f(x) \qquad x \in \mathbb{R}^m \tag{3.1}$$

with an *isolated*[1] fixed point $\bar{x}$ such that $f(\bar{x}) = 0$. Let $x(t)$ denote the state of the system at time t, $x(0) = x_0$ the initial point and $\| \cdot \|$ indicate the

---

[1] A fixed point is **isolated** if it has a surrounding neighbourhood containing no other fixed point.

Euclidean distance in $\mathbb{R}^m$.[2] With this notation we can now introduce the following definitions of stability.

definition 3.1 *The fixed point $\bar{x}$ is said to be* **Lyapunov stable** *(or simply stable) if for any $\epsilon > 0$ there exists a number $\delta(\epsilon) > 0$ such that if $\|x_0 - \bar{x}\| < \delta$ then $\|x(t) - \bar{x}\| < \epsilon$ for all $t > 0$.*

remark 3.1 Important aspects of stability in the sense of Lyapunov are:

(1) the constraint on the initial condition is that $\delta(\epsilon)$ must exist for *any* $\epsilon > 0$;

(2) for the generality of nonautonomous systems of differential equations, the number δ depends also on the initial time t_0, whereas, for autonomous systems δ depends only on ϵ. In this case we say that stability is uniform. Uniformity also holds for nonautonomous systems which are periodic in the sense that we have $\dot{x} = f(t, x) = f(t + \omega, x)$ for some ω.

definition 3.2 *The fixed point $\bar{x}$ is said to be* **asymptotically stable** *if*

(a) it is stable; and

(b) there exists an $\eta > 0$ such that whenever $\|x_0 - \bar{x}\| < \eta$

$$\lim_{t \to \infty} \|x(t) - \bar{x}\| = 0.$$

Lyapunov and asymptotic stability are represented in figure 3.1.

Because system (3.1) is autonomous, asymptotic stability is uniform and property *(b)* of definition 3.2 can be replaced by the equivalent property

(b') there exists a number $\eta > 0$ and, for each $\epsilon > 0$, a real positive number $T = T(\eta, \epsilon)$, such that if $\|x_0 - \bar{x}\| < \eta$ then $\|x(t) - \bar{x}\| < \epsilon$ for all $t > T$.

Notice that because of uniformity, T does not depend on initial time but only on η and ϵ.

definition 3.3 *Let $\bar{x}$ be an asymptotically stable fixed point of (3.1), then the set*

$$B(\bar{x}) = \{x \in \mathbb{R}^m \mid \lim_{t \to \infty} \|x(t) - \bar{x}\| = 0\}$$

---

[2] Recall that the Euclidean distance between 2 points $x = (x_1, x_2, \ldots, x_m)^T$ and $y = (y_1, y_2, \ldots, y_m)^T$ in $\mathbb{R}^m$ is equal to $\sqrt{\sum_{i=1}^{m} (x_i - y_i)^2}$.

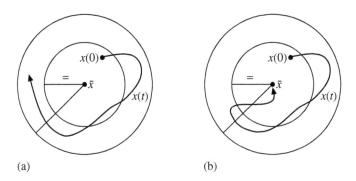

Fig. 3.1 (a) Lyapunov stability; (b) asymptotic stability

is the **domain** *or* **basin of attraction** *of* $\bar{x}$*. If* $B(\bar{x}) = \mathbb{R}^m$ *(or, at any rate, if it coincides with the state space) then* $\bar{x}$ *is said to be* **globally asymptotically stable***. If stability only holds in a neighbourhood of* $\bar{x}$,[3] *it is said to be* **locally stable***.*

Broadly speaking, when defined in terms of the union of properties *(a)* and *(b)*, asymptotic stability implies that if orbits start near equilibrium they stay near it and eventually converge to it. The equivalent formulation *(b′)* indicates that, starting sufficiently near equilibrium, any arbitrarily small neighbourhood of it is reached in finite time.

remark 3.2 Property *(b)*, or *(b′)*, (convergence to equilibrium) does not imply property *(a)* (stability). That is, if *(b)*, or *(b′)*, holds but *(a)* does not, we could have solutions that, before converging to $\bar{x}$, wander arbitrarily far from it. (Convergence does not imply that the maximum distance from equilibrium goes to zero with the initial distance.)

Systems for which property *(a)* holds, but property *(b)*, or *(b′)*, does not, are called **weakly stable**. We have seen an example of this type of stability in chapter 2 in the case of centres. Examples of asymptotic stability were given by stable nodes or foci. Systems for which property *(b)*, or *(b′)*, holds, but property *(a)* does not, are sometimes called **quasi-asymptotically stable**. However, for *linear systems*, property *(b)*, or *(b′)*, does imply property *(a)*.

---

[3]Let $A \subset M$ be a nonempty subset of a metric space M. We say that the set N is a **neighbourhood** of A if N contains an open set containing A. For example the sphere $N(\epsilon, A) = \{x \in M | d(x, A) < \epsilon\}$ is an open neighbourhood of A and $N(\epsilon, A) = \{x \in M | d(x, A) \leq \epsilon\}$ is a closed neighbourhood of A. In this case we sometimes use the expression ϵ-neighbourhood and, when necessary, specify if N is open or closed.

remark 3.3 A better understanding of uniformity in asymptotically stable systems can be had by comparing two simple differential equations. Consider, first

$$\dot{x} = -x \qquad x \in \mathbb{R} \tag{3.2}$$

which is an autonomous differential equation with a unique equilibrium $\bar{x} = 0$ and solution

$$x(t) = x_0 e^{-(t-t_0)} \qquad x_0 = x(t_0).$$

The equilibrium is asymptotically stable and in particular, it is true that

$$\lim_{t \to \infty} |x(t)| = 0 \qquad \forall x_0 \in \mathbb{R}.$$

Moreover, starting from any point x_0, with distance $|x_0|$ from equilibrium, the system will reach a point $x(t)$ with distance ϵ from equilibrium in a time $T \equiv t - t_0 > 0$ equal to

$$T = \ln \frac{|x_0|}{\epsilon}.$$

Therefore, for any given value of $|x_0|$ and ϵ, T is independent of the choice of t_0 and asymptotic stability is uniform.

The second equation

$$\dot{x} = -\frac{x}{1+t} \qquad x \in \mathbb{R} \tag{3.3}$$

is a nonautonomous differential equation with unique equilibrium at $\bar{x} = 0$ and solution

$$x(t) = x_0 \frac{1+t_0}{1+t}.$$

Again, we have

$$\lim_{t \to \infty} |x(t)| = 0 \qquad \forall x_0 \in \mathbb{R}.$$

However the time interval $T \equiv t - t_0 > 0$ required to move from a point x_0, with distance $|x_0|$ from the equilibrium point, to a point $x(t)$ with distance ϵ, is equal to

$$T = \left(\frac{|x_0|}{\epsilon} - 1 \right)(1 + t_0).$$

Therefore, T depends on t_0 as well as on $|x_0|$ and ϵ and asymptotic stability is not uniform.

There exist analogous definitions of stability for autonomous dynamical systems in discrete time with the general form

$$x_{n+1} = G(x_n) \qquad x \in \mathbb{R}^m. \tag{3.4}$$

Suppose $\bar{x}$ is an isolated equilibrium point of (3.4), such that $G(\bar{x}) = \bar{x}$.

definition 3.4 *The equilibrium point $\bar{x}$ is* **Lyapunov stable** *(or, simply,* **stable***) if for every $\epsilon > 0$, there exists $\delta(\epsilon)$ such that*

$$[\|x_0 - \bar{x}\| < \delta(\epsilon)] \Rightarrow [\|G^n(x_0) - \bar{x}\| < \epsilon \ \forall n > 0].$$

definition 3.5 *The equilibrium point $\bar{x}$ is* **asymptotically stable** *if*

(a) *it is stable* and
(b) $\exists\, \eta > 0$ *such that*

$$[\|x_0 - \bar{x}\| < \eta] \Rightarrow \left[\lim_{n \to \infty} \|G^n(x_0) - \bar{x}\| = 0 \right].$$

Property *(b)* can be replaced by the equivalent property

(b') *there exists $\eta > 0$ and, for each $\epsilon > 0$, there exists an integer $T = T(\eta, \epsilon) > 0$ such that*

$$[\|x_0 - \bar{x}\| < \eta] \Rightarrow [\|G^n(x_0) - \bar{x}\| < \epsilon \ \forall n \geq T].$$

Notice that for autonomous systems like (3.4), stability and asymptotic stability are uniform and δ and T do not depend on the initial time. Remark 3.2 applies to discrete-time systems too. For certain systems, asymptotical stability may take a stronger form as defined below.

definition 3.6 *An equilibrium $\bar{x}$ of system (3.1) is said to be* **exponentially stable** *if there exist two positive constants α and β such that, if $\|x_0 - \bar{x}\| < \eta$ for some $\eta > 0$, then we have*

$$\|x(t) - \bar{x}\| < \beta \|x_0 - \bar{x}\| e^{-\alpha t} \ \forall t > 0.$$

The constants α and β may depend on η.

An entirely analogous definition can be written for a discrete-time system like (3.4).

Asymptotically stable *linear* systems (in both continuous and discrete time), such as those studied in chapter 2, are all exponentially stable. It should be observed that the properties *stable* and *unstable* are not symmetric in the sense that stability is defined in terms of the behaviour of a family of orbits, whereas instability occurs whenever even a single orbit violates the conditions for stability.

definition 3.7 *A fixed point of systems (3.1) or (3.4) is called* **unstable** *if it is not stable.*

This means that there exists a number $\epsilon > 0$ and an orbit starting from a point x_0 arbitrarily close to the fixed point such that the distance between the orbit and the fixed point becomes at least as large as ϵ at a future moment, that is, $\|x(t) - \bar{x}\| \geq \epsilon$ for some $t > 0$. The unstable nodes, unstable foci, and saddle points discussed for linear systems in chapter 2 are examples of unstable fixed points according to the definition given above. We have an entirely analogous definition for discrete-time systems.

remark 3.4 In general, linear systems have a unique equilibrium point which is located, in the homogeneous case, at the origin. For those systems, asymptotic stability is always global (or 'in the large'), that is to say, it does not depend on the distance from the equilibrium; moreover asymptotic stability is always exponential. Weakly stable systems are exceptional (for example, the centre as discussed in chapter 2, section 2.3). Nonlinear systems, on the other hand, may have any finite, or a countably infinite number of fixed points; local asymptotic stability does not necessarily imply global stability, nor is it necessarily exponential; weak stability is not exceptional.

3.2 The linear approximation

In chapter 1 we mentioned that for the generality of nonlinear dynamical systems, closed form solutions are not available. Therefore, stability of those systems usually must be determined by methods that do not require the knowledge of their solutions. We begin with the method based on the linear approximation.

Consider again the nonlinear system of differential equations of system (3.1) $\dot{x} = f(x)$, $x \in \mathbb{R}^m$ with a fixed point $\bar{x}$. Assume, as we usually do, that f is differentiable in each of the m variables of the vector x. A local linear approximation of the nonlinear system near the fixed point $\bar{x}$ is given by the expansion in a Taylor series of the coordinate functions f_i (the elements of the vector f) truncated after the first-order terms. Let $\xi = x - \bar{x}$, whence

$\dot{\xi} = \dot{x}$. Then we can write (in first approximation)

$$\dot{\xi}_1 = f_1(\bar{x}) + \frac{\partial f_1(\bar{x})}{\partial x_1}\xi_1 + \frac{\partial f_1(\bar{x})}{\partial x_2}\xi_2 + \cdots + \frac{\partial f_1(\bar{x})}{\partial x_m}\xi_m$$

$$\dot{\xi}_2 = f_2(\bar{x}) + \frac{\partial f_2(\bar{x})}{\partial x_1}\xi_1 + \frac{\partial f_2(\bar{x})}{\partial x_2}\xi_2 + \cdots + \frac{\partial f_2(\bar{x})}{\partial x_m}\xi_m$$

$$\vdots \quad \vdots \tag{3.5}$$

$$\dot{\xi}_m = f_m(\bar{x}) + \frac{\partial f_m(\bar{x})}{\partial x_1}\xi_1 + \frac{\partial f_m(\bar{x})}{\partial x_2}\xi_2 + \cdots + \frac{\partial f_m(\bar{x})}{\partial x_m}\xi_m.$$

Considering that the first term on the RHS of each equation in system (3.5) is equal to zero, the linearised system can be written compactly as

$$\dot{\xi} = Df(\bar{x})\xi \tag{3.6}$$

where, in general, $Df(x)$ denotes the matrix of first partial derivatives (often called 'Jacobian' matrix), namely

$$Df(x) = \begin{pmatrix} \frac{\partial f_1}{\partial x_1} & \frac{\partial f_1}{\partial x_2} & \cdots & \frac{\partial f_1}{\partial x_m} \\ \frac{\partial f_2}{\partial x_1} & \frac{\partial f_2}{\partial x_2} & \cdots & \frac{\partial f_2}{\partial x_m} \\ \vdots & \vdots & \ddots & \vdots \\ \frac{\partial f_m}{\partial x_1} & \frac{\partial f_m}{\partial x_2} & \cdots & \frac{\partial f_m}{\partial x_m} \end{pmatrix}.$$

It is important to know how much of the dynamic characterisation of the original nonlinear system (3.1) is preserved by the approximation (3.6). There exists a general result in the theory of differential equations known as the Hartman–Grobman theorem, which guarantees that, under appropriate conditions, in a neighbourhood of the equilibrium point the qualitative properties of the nonlinear system are preserved by the linearisation. The conditions regard the hyperbolicity of the fixed point.

definition 3.8 *A fixed point of a system of differential equations is* **hyperbolic** *if the Jacobian matrix calculated at that point has no zero or purely imaginary eigenvalues (no eigenvalue has real part equal to zero).*

Now we can state the following fundamental theorem (see Hartman, 1964, p. 244, theorem 7.1; Guckenheimer and Holmes, 1983, p. 13, theorem 1.31).

theorem 3.1 (Hartman–Grobman) *If $\bar{x}$ is a hyperbolic fixed point of (3.1), then there is a homeomorphism[4] h defined on some neighbourhood N of $\bar{x}$ in $\mathbb{R}^m$, locally taking orbits of the nonlinear system (3.1) to those of*

---

[4] A continuous map g that is one-to-one (injective) and onto (surjective), and therefore invertible, with the property that both g and g^{-1} are k times differentiable, is called a C^k **diffeomorphism**. If $k = 0$ the map is called a **homeomorphism**.

the linear system (3.6). The map h preserves the sense of orbits and can also be chosen so as to preserve parametrisation by time.

If h is a homeomorphism, then from theorem 3.1 we can deduct that asymptotic stability (or the lack of it) for the linear system (3.6) implies local asymptotic stability of the nonlinear system (3.1) (or the lack of it). However, homeomorphic equivalence does not preserve all the interesting geometric features of a dynamical system. For example, a linear system characterised by an asymptotically stable node is topologically conjugate (there is more on this type of equivalence in chapter 6) to another linear system characterised by an asymptotically stable focus.

The equivalence between a nonlinear system such as (3.1) and its linearisation (3.6) would be stronger if h were a diffeomorphism. This has been proved to be the case for the nonlinear system (3.1) whenever the eigenvalues of the corresponding matrix $Df(\bar{x})$ satisfy a 'nonresonance condition'. This condition requires that, for any choice of coefficients $c_i \geq 0$ with $\sum_{i=1}^{m} c_i \geq 2$, and for any eigenvalue λ_k of $DF(\bar{x})$, $\lambda_k \neq \sum_{i=1}^{m} c_i \lambda_i$.

If the equilibrium point is not hyperbolic, that is to say, if there exists at least one eigenvalue with real part exactly equal to 0, the Hartman–Grobman theorem cannot be applied. The reason is that the linearised system is not sufficiently informative. In particular, the stability properties of the system depend on the higher-degree terms of the expansion which have been ignored in the approximations (3.5) and (3.6).

In our discussion of linear systems we emphasised the importance of certain invariant subspaces, the eigenspaces, defined by the eigenvectors of the controlling matrix. If the nonlinear system (3.1) has an isolated, hyperbolic equilibrium $\bar{x}$, in the neighbourhood of $\bar{x}$ there exist certain invariant surfaces, called stable and unstable manifolds, which are the nonlinear counterparts of the stable and unstable eigenspaces.[5] Locally, these manifolds are continuous deformations, respectively, of the stable and unstable eigenspaces of the linear system (3.1) (because $\bar{x}$ is hyperbolic, there is no centre eigenspace for (3.1)) and they are tangent to them at $\bar{x}$.

Some simple examples of the phase diagrams of nonlinear systems and the corresponding linearised systems in $\mathbb{R}^2$ and $\mathbb{R}^3$ are provided in figures 3.2, 3.3 and 3.4.

The method of linear approximation can be applied in a perfectly analogous manner to nonlinear systems of difference equations. Consider system (3.4), with a fixed point $\bar{x}$, and assume that G is differentiable in the m variables of the vector x. A local linear approximation of (3.4) near $\bar{x}$ is

---

[5]We refer the reader to appendix C, p. 98, for a more precise definition of a manifold.

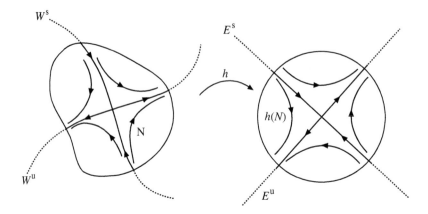

Fig. 3.2 The Hartman–Grobman theorem

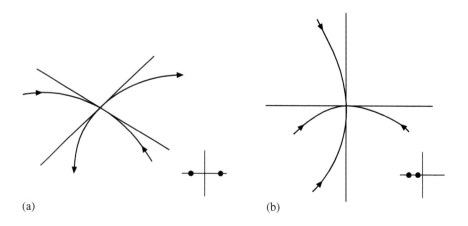

(a)   (b)

Fig. 3.3 Stable and unstable eigenspaces and manifolds

again given by truncated Taylor expansions of the functions G_i. Letting
$\xi = x - \bar{x}$, we have (in first approximation)

$$x_{n+1}^{(1)} = \xi_{n+1}^{(1)} + \bar{x}^{(1)} = G_1(\bar{x}) + \frac{\partial G_1(\bar{x})}{\partial x_n^{(1)}}\xi_n^{(1)} + \cdots + \frac{\partial G_1(\bar{x})}{\partial x_n^{(m)}}\xi_n^{(m)}$$

$$x_{n+1}^{(2)} = \xi_{n+1}^{(2)} + \bar{x}^{(2)} = G_2(\bar{x}) + \frac{\partial G_2(\bar{x})}{\partial x_n^{(1)}}\xi_n^{(1)} + \cdots + \frac{\partial G_2(\bar{x})}{\partial x_n^{(m)}}\xi_n^{(m)}$$

$$\vdots \qquad \vdots \qquad\qquad \vdots$$

$$x_{n+1}^{(m)} = \xi_{n+1}^{(m)} + \bar{x}^{(m)} = G_m(\bar{x}) + \frac{\partial G_m(\bar{x})}{\partial x_n^{(1)}}\xi_n^{(1)} + \cdots + \frac{\partial G_m(\bar{x})}{\partial x_n^{(m)}}\xi_n^{(m)}$$

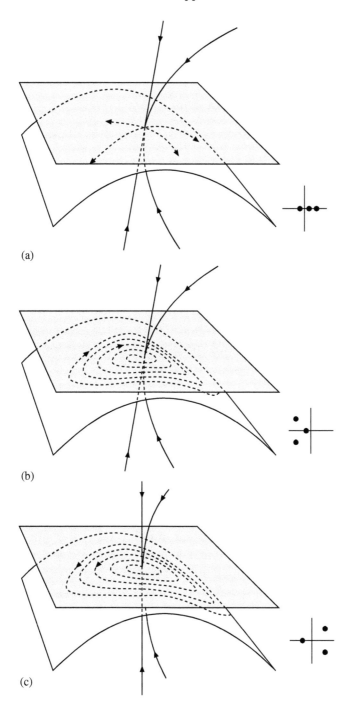

(a)

(b)

(c)

Fig. 3.4 Stable and unstable eigenspaces and manifolds in $\mathbb{R}^3$

where $x_n^{(i)}$, $\xi_n^{(i)}$, $\bar{x}^{(i)}$ denote, respectively, the ith element of the vectors x_n, ξ_n, $\bar{x}$ and G_i is the ith coordinate function of G.

In equilibrium, $\bar{x}^{(i)} = G_i(\bar{x})$ for all i and therefore the linearised system can be written compactly as

$$\xi_{n+1} = DG(\bar{x})\xi_n \tag{3.7}$$

where $DG(\bar{x})$ is the Jacobian matrix of partial derivatives of G, evaluated at $\bar{x}$.

The version of the Hartman–Grobman theorem for diffeomorphisms of the form $x_{n+1} = G(x_n)$ is perfectly analogous to that for flows[6] and we wish to stress only the following important differences.

(i) For discrete-time systems, fixed points are hyperbolic if none of the eigenvalues of the Jacobian matrix, evaluated at the equilibrium, is *equal to 1 in modulus.*

(ii) The map h of the Hartman–Grobman theorem defining the local relation between the nonlinear system (3.4) and the linearised system (3.7) is a diffeomorphism if the eigenvalues of $DG(\bar{x})$ satisfy a nonresonance condition. In the case of maps this condition requires that for no eigenvalue κ_k of $DG(\bar{x})$ we have

$$\kappa_k = \prod_{i=1}^{m} \kappa_i^{c_i}$$

for any choice of $c_i \geq 0$ with $\sum_i c_i \geq 2$.

3.3 The second or direct method of Lyapunov

The so-called **second** or **direct method** of Lyapunov is one of the greatest landmarks of the theory of dynamical systems and has proved an immensely fruitful tool of analysis. There also exists a *first method* of Lyapunov, including mathematical techniques of analysis of dynamical systems and their stability which make use of the solutions of those systems in explicit form, especially as infinite series (see Hartman, 1964, pp. 337–41). We mention it here only as a matter of historical curiosity and shall not use it again. The great insight of the second or direct method of Lyapunov is precisely that general statements about stability of an equilibrium point of a system can be made without any prior knowledge of its solution. As we shall see later, the method can be employed also in the analysis of stability of sets 'larger' than a point.

[6] See Hartman (1964), p. 245, lemma 8.1.

The method makes use of certain functions (typically of a distance-like form) and attempts to determine stability by evaluating the sign of their time derivatives along orbits of the system. In some cases, simple illuminating geometrical interpretations are possible.

The following is the core statement of Lyapunov's direct method.

theorem 3.2 *Consider the system of differential equations (3.1)*

$$\dot{x} = f(x) \qquad x \in \mathbb{R}^m.$$

Let $\bar{x}$ be an isolated equilibrium point which we assume to be at the origin. If there exists a C^1 scalar function $V(x)\colon N \to \mathbb{R}$, defined on some neighbourhood $N \subset \mathbb{R}^m$ of 0, such that

(i) $V(0) = 0$;
(ii) $V(x) > 0$ *in* $N \setminus \{0\}$;
(iii) $\dot{V}(x) = \sum_{i=1}^m \frac{\partial V}{\partial x_i} f_i(x) \leq 0$ *in* $N \setminus \{0\}$;

then $\bar{x} = 0$ is stable (in the sense of Lyapunov). Moreover, if

(iv) $\dot{V}(x) < 0$ *in* $N \setminus \{0\}$

then $\bar{x} = 0$ is asymptotically stable.

Functions satisfying (i)–(iii) or (i)–(iv) are often called **Lyapunov functions**.

It is sometimes possible to prove *asymptotical* stability of a fixed point even when the Lyapunov function V in the relevant neighbourhood of the point implies $\dot{V} \leq 0$, but not necessarily $\dot{V} < 0$. To that end we invoke the following result by Krasovskii and La Salle (see Krasovskii, 1963, p. 67, theorem 14.1; Glendinning, 1994, p. 37, theorem 2.10, on which the following statement of the result is based).

theorem 3.3 *Let $\bar{x} = 0$ be a fixed point of $\dot{x} = f(x)$ and V a Lyapunov function such that $\dot{V}(x) \leq 0$ on some neighbourhood N of $\bar{x} = 0$. Let $\phi(t, x)$ denote the flow map generated by f. If $x_0 \in N$ has its forward orbit, $\gamma^+(x_0) = \{\phi(t, x_0)|t \geq 0\}$, bounded with limit points in N, and M is the largest invariant subset of $E = \{x \in N \mid \dot{V}(x) = 0\}$, then*

$$\phi(t, x_0) \to M \qquad as \ t \to \infty.$$

Thus, if a Lyapunov function $V(x)$ can be found such that $\dot{V}(x) \leq 0$ for $x \in N$, among the sets of points with forward orbits in N there exist sets of points defined by

$$V_k = \{x \mid V(x) \leq k\}$$

(k a finite, positive scalar) which lie entirely in N. Because $\dot{V} \leq 0$, the sets V_k are invariant in the sense that no orbit starting in a V_k can ever move outside it. If, in addition, it could be shown that the fixed point $\bar{x} = 0$ is the largest (or, for that matter, the only) invariant subset of E, theorem 3.3 would guarantee its asymptotic stability.[7]

The direct method can also be extended to discrete-time systems as follows.

theorem 3.4 *Consider the system of difference equations (3.4)*

$$x_{n+1} = G(x_n), \qquad x \in \mathbb{R}^m.$$

Let $\bar{x}$ again be an isolated equilibrium point at the origin. If there exists a C^1 function $V(x_n)$: $N \to \mathbb{R}$, defined on some neighbourhood $N \subset \mathbb{R}^m$ of 0, such that

(i) $V(0) = 0$;
(ii) $V(x_n) > 0$ in $N \setminus \{0\}$;
(iii) $\Delta V(x_n) \equiv V[G(x_n)] - V(x_n) \leq 0$ in $N \setminus \{0\}$;

then $\bar{x} = 0$ is stable (in the sense of Lyapunov). If, moreover,

(iv) $\Delta V(x_n) < 0$ in $N \setminus \{0\}$

then $\bar{x} = 0$ is asymptotically stable.

Lyapunov-type functions have also been used (by Lyapunov himself as well as other authors) to prove *instability* of equilibrium points. For example, we have the following so-called **first Lyapunov theorem on instability**, for systems of differential equations.

theorem 3.5 *Consider system (3.1) and let there be a fixed point $\bar{x} = 0$. Suppose there exists a continuously differentiable scalar function $V : N \to \mathbb{R}$, defined on some neighbourhood N of 0 and such that*

(i) $V(0) = 0$;
(ii) $\dot{V}(x)$ is positive definite.

Then if $\bar{x} = 0$ is an accumulation point[8] of the set of points for which $V > 0$ (in other words, $V(x)$ can take positive value arbitrarily near $\bar{x}$), $\bar{x}$ is unstable.

---

[7] Theorem 3.3 is sometimes called the 'Invariance Principle of La Salle' and can be extended to the case in which $\dot{x} = f(t, x)$ with f periodic in time. For more general results that prove asymptotic stability requiring that the time derivative of the Lyapunov function be only negative *semi*definite, see Bhatia and Szegö (1970), pp. 156–62.

[8] Let X be a metric space. A point $x \in X$ is said to be an **accumulation point** of a subset $A \subset X$ if in every neighbourhood N of x there lie points of the set $A \setminus \{x\}$.

The essential step in practical applications of Lyapunov's second method is the definition of a function $V(x)$ with the required properties. Unfortunately, there are no ready-made methods for finding such a function and much is left to the intuition and experience of the investigator. A good guess is to try some sort of distance function or a sign-preserving transformation of it, for these will automatically satisfy (ii). Then if the fixed point is at the origin (which can always be accomplished by an appropriate coordinate change), condition (i) is also satisfied. A distance-like function is an intuitively appealing choice because if the equilibrium of the system is asymptotically stable, the distance from it must eventually decline and indeed, go to zero. A geometric interpretation of the Lyapunov method can be given by a very simple two-dimensional case:

$$\dot{x} = y$$
$$\dot{y} = -x - y.$$

The equilibrium point is $(0,0)$ and a Lyapunov function is

$$V(x, y) = \frac{1}{2}(x^2 + y^2)$$

with $\dot{V}(x, y) = x\dot{x} + y\dot{y} = -y^2 < 0$ for $y \neq 0$. To different given positive values of $V(x, y) = k_i$ there correspond circles[9] around the origin with radius $= \sqrt{2k_i}$. The fact that $\dot{V}(x, y) < 0$ along the orbits clearly indicates that the state of the system moves in time towards smaller and smaller circles, asymptotically converging to the equilibrium at the origin (see figure 3.5).

However, in general, there is no guarantee (even for asymptotically stable systems) that distance should be *monotonically* declining in a neighbourhood of the equilibrium (see, for example, figure 3.6).

For a certain class of two-dimensional systems of differential equations, however, a suitable candidate for a Lyapunov function can be defined by means of a simple and useful *rule of thumb*. The system must have the following general form (or it must be possible to manipulate it into the form)

$$\dot{x} = -y - f(x) \tag{3.8}$$
$$\dot{y} = g(x). \tag{3.9}$$

Differentiating (3.8) with respect to time and using (3.9) we have

$$\ddot{x} + f'(x)\dot{x} + g(x) = 0.$$

[9] These closed curves on each of which the value of a Lyapunov function is constant are called **contour lines** or **level curves**, as they suggest the curves on topographical maps.

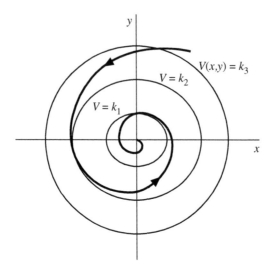

Fig. 3.5 Contour lines of V, trajectories of $\dot{x} = f(x)$ cross with inward direction

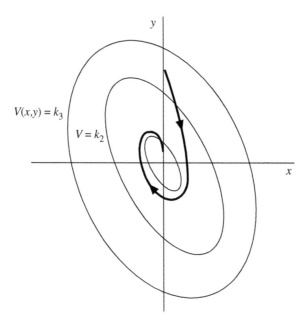

Fig. 3.6 The distance function decreases asymptotically not monotonically with time

If we multiply this equation by $\dot{x}$ we get

$$\dot{x}\ddot{x} + \dot{x}^2 f'(x) + \dot{x}g(x) = 0$$

or

$$\frac{d}{dt}\left(\frac{1}{2}\dot{x}^2 + \int_0^x g(s)ds\right) = -\dot{x}^2 f'(x).$$

Suppose now that $\bar{x}$ is a fixed point of (3.8)–(3.9). We could try the putative Lyapunov function

$$V(x) = \left(\frac{1}{2}\dot{x}^2 + \int_0^x g(s)ds\right) \tag{3.10}$$

whence

$$\dot{V}(x) = -\dot{x}^2 f'(x) \tag{3.11}$$

and verify under what constraints the functions (3.10) and (3.11) satisfy the conditions in theorem 3.2 (or theorem 3.3). For example, if $f'(x) > 0$ in a certain neighbourhood N of $\bar{x}$, then (3.11) implies that $\dot{V}(x) \le 0$ for $x \in N$. This establishes stability of $\bar{x}$. If, moreover, the conditions of theorem 3.3 hold, then $\bar{x}$ is asymptotically stable.

Notice that in theorems 3.2 and 3.4 the existence of a function with the required properties is a *sufficient* condition for stability (or, respectively, asymptotic stability) of equilibria of systems of differential or difference equations. The theorems themselves do not prove that such a function exists. Fortunately, there exist associated results (by Lyapunov and several other authors) proving that the conditions are also *necessary*. Such theorems are often known as **converse theorems** and similar theorems also exist for difference equations. We return to converse theorems in chapter 4 when we discuss stability of general sets. Here we only present a result originally proved by Lyapunov and regarding the simpler linear case. Consider the system

$$\dot{x} = Ax \qquad x \in \mathbb{R}^m$$

where A is a $(m \times m)$ constant matrix. If A is nonsingular, $\bar{x} = 0$ is the unique equilibrium point. Suppose now all of the eigenvalues of A have negative real parts and therefore $\bar{x} = 0$ is (globally) asymptotically stable.

Then Lyapunov's converse theorem states that there exists a symmetric, positive-definite matrix $B = B^T$ such that the function

$$V(x) = x^T B x \tag{3.12}$$

satisfies all the properties of theorem 3.2 with $\dot{V}(x) < 0$. Properties (i) and (ii) are satisfied trivially. As concerns property (iii), consider that

$$\dot{V}(x) = x^T B \dot{x} + \dot{x}^T B x$$
$$= x^T (BA + A^T B)x$$

where as usual ^{T} denotes transposition. Property (iii) is therefore satisfied
if

$$A^T B + BA = -C \tag{3.13}$$

with C any symmetric, positive-definite matrix. This will be the case if we
choose

$$B = \int_0^\infty e^{A^T t} C e^{At} dt.$$

To see this, consider that B is positive-definite wherever C is and the integral
converges under the postulated assumptions on A. Substituting this integral
into (3.13) we have

$$A^T \int_0^\infty e^{A^T t} C e^{At} dt + \int_0^\infty e^{A^T t} C e^{At} dt \, A$$

$$= \int_0^\infty \left(A^T e^{A^T t} C e^{At} + e^{A^T t} C e^{At} A \right) dt$$

$$= \int_0^\infty \frac{d}{dt} \left(e^{A^T t} C e^{At} \right) dt$$

$$= e^{A^T t} C e^{At} \Big|_0^\infty = -C.$$

We shall return to other, more general converse theorems when we discuss
the stability of sets in chapter 4.

Lyapunov functions can also be used to prove, in a relatively simple manner, a result that we have mentioned before, stating that asymptotic stability of the equilibrium point of a linearised system implies asymptotic
stability of the original nonlinear system in a neighbourhood of equilibrium.
Take the nonlinear system (3.1)

$$\dot{x} = f(x) \qquad x \in \mathbb{R}^m$$

and assume that there exists a locally unique equilibrium point $\bar{x} = 0$. If f
is differentiable, (3.1) can be written as

$$\dot{x} = Ax + h(x) \tag{3.14}$$

where $A = Df(\bar{x})$ is the Jacobian matrix of partial derivatives evaluated at
$\bar{x} = 0$, and $h(x)$ contains terms of second and higher order. Suppose now
that all eigenvalues of A have negative real parts. Then, as we have just
shown, there exists a positive-definite matrix B such that, if we choose

$$V_L(x) = x^T Bx,$$

the total derivative of $V_L(x)$ with respect to time, calculated along orbits of the linearised system $\dot{x} = Ax$ is

$$\dot{V}_L(x) = x^T B\dot{x} + \dot{x}^T Bx = x^T BAx + x^T A^T Bx = -x^T Cx < 0. \qquad (3.15)$$

Consider now the complete nonlinear system (3.14) and choose $V_{NL}(x) = x^T Bx$ as a Lyapunov function for the nonlinear system. The total derivative of $V_{NL}(x)$ along orbits of (3.14) is

$$\dot{V}_{NL}(x) = \dot{V}_L(x) + x^T Bh(x) + h^T(x)Bx.$$

From (3.15) and the symmetry of B we have

$$\dot{V}_{NL}(x) = -x^T Cx + 2x^T Bh(x). \qquad (3.16)$$

On the RHS of (3.16) the quadratic form $-x^T Cx$ is negative (for $x \neq 0$), and contains terms of second degree in x, whereas the sign of $2x^T Bh(x)$ is unknown, but we know that it contains terms of third and higher degree in x. Therefore, if we choose $\|x\|$ small enough (that is, we consider orbits starting sufficiently near the equilibrium point), the negative terms in (3.16) dominate and we have

$$\dot{V}_{NL}(x) < 0.$$

Thus, locally, the (asymptotic) stability of the linearised system carries over to the complete nonlinear system.

remark 3.5 If the equilibrium point of the linearised system $\dot{x} = Ax$ is stable but not asymptotically stable (that is, there are eigenvalues of A with zero real parts and the equilibrium is not hyperbolic), we cannot draw any definite conclusions from the analysis of that system about the stability of the complete nonlinear system because, in this case, the nonlinear part $h(x)$ of (3.14) and (3.16) plays an essential role in determining the stability properties of the equilibrium point.

Quadratic forms are a natural choice for Lyapunov functions of linear systems but certain generalisations of them are used for nonlinear systems as well. We shall consider here some particularly interesting quadratic forms employed in applications, including economic dynamics.[10]
Consider again the system

$$\dot{x} = Ax$$

as the linearisation, calculated at the equilibrium point $\bar{x} = 0$, of the general autonomous system of differential equations (3.1)

$$\dot{x} = f(x).$$

[10] See Hartman (1964), pp. 539–40; Brock and Malliaris (1989), chapter 4, pp. 89–129.

$A = Df(\bar{x})$ is the constant matrix of partial derivatives calculated at $\bar{x} = 0$. The results with which we are concerned here are based on certain assumed properties of the *variable* matrix $Df(x)$, evaluated along the orbits of the system.

To investigate global stability of $\bar{x}$ we try again the function (3.12)

$$V(x) = x^T B x$$

where B is a symmetric, positive-definite constant matrix. Then $V(x) \geq 0$ if $\|x\| \geq 0$, and $V(x) \to \infty$ as $\|x\| \to \infty$. Taking the total derivative of $V(x)$ with respect to time along orbits of the system we have

$$\dot{V}(x) = x^T B \dot{x} + \dot{x}^T B x = 2\dot{x}^T B x$$
$$= 2 f(x)^T B x. \tag{3.17}$$

To prove global stability and establish the conditions for which $\dot{V}(x) \leq 0$ if $\|x\| \geq 0$, we proceed as follows. Let s be a scalar $s \in [0,1]$ and define the vector $z = sx$. Then

$$f(x) = \int_0^1 [Df(z)] \frac{\partial z}{\partial s} ds = \int_0^1 [Df(z)] x \, ds \tag{3.18}$$

where $[Df(z)]$ is the matrix of partial derivatives of f with respect to the elements of the vector z. Consequently, we have

$$f(x)^T B x = \int_0^1 x^T [Df(z)]^T B x \, ds$$
$$= \int_0^1 x^T B [Df(z)] x \, ds. \tag{3.19}$$

Then, if we can establish that, for some symmetric, positive-definite matrix B, the quadratic form $x^T B [Df(z)] x < 0$ for all $x \neq 0$ and all $z \neq 0$, we can conclude that the fixed point $\bar{x} = 0$ of system (3.1) is globally asymptotically stable. Alternatively, suppose $\|f(x)\| \to \infty$ as $\|x\| \to \infty$ and choose

$$V(x) = \dot{x}^T B \dot{x} = f(x)^T B f(x) \tag{3.20}$$

with B again a symmetric, positive-definite constant matrix. In this case we have $V(x) \geq 0$ if $\|x\| \geq 0$ and $V(x) \to \infty$ if $\|x\| \to \infty$. Also, differentiating with respect to time we have

$$\dot{V}(x) = f(x)^T [Df(x)]^T B f(x) + f(x)^T B [Df(x)] f(x)$$
$$= 2 f(x)^T B [Df(x)] f(x). \tag{3.21}$$

Then, if for some symmetric positive-definite matrix B, the quadratic form

$x^T B[Df(x)]x < 0$ for all $x \neq 0$, the fixed point $\bar{x} = 0$ is globally, asymptotically stable.

remark 3.6 The property $\|f(x)\| \to \infty$ as $\|x\| \to \infty$, is sometimes summarised by calling the function $f(x)$ **radially unbounded**, and is introduced to guarantee that $f(x)$ and thereby $V(x)$ in (3.20) are defined globally. That requirement is automatically satisfied for $V(x)$ as in (3.12).

Other global stability conditions on the Jacobian matrix, sometimes used in applications, including economics, are the following:

$$y^T [Df(x)]y \leq 0 \qquad \text{if } f(x)^T y = 0$$

or

$$y^T B[Df(x)]y \leq 0 \qquad \text{if } f(x)^T By = 0$$

where f, $Df(x)$ and B are defined as before.

There also exist generalisations of the stability conditions discussed above in which the constant matrix B is replaced by a matrix function $B(x)$ (for details see Hartman, 1964, pp. 548–55).

remark 3.7 Lyapunov's and other similar methods are most often applied to the study of stability of equilibria. However, we are sometimes interested not so much in proving that orbits of a system converge to a point (or to a set) as in ensuring that, even if the equilibrium point is unstable, orbits starting in a certain reasonable set of initial conditions remain bounded. This may sometimes be obtained by establishing the existence of a trapping region, that is, a region such that all orbits starting on its boundary, point towards its interior (cf. definition 4.6 and in point (iii) on p. 226). In some cases this can be obtained by defining a Lyapunov-type function whose total derivative with respect to time along the orbits of the system is negative for states of the system sufficiently far from the fixed point. The existence of trapping regions can be similarly proved in systems of difference equations.

Appendix A: general economic equilibrium

In this and the following appendix B we discuss two classical problems of economic dynamics which have been studied by means of the stability methods discussed in the previous sections. Here, we consider the question of stability of a pure exchange, competitive equilibrium with an adjustment mechanism known as *tâtonnement* and directly inspired by the work of Léon Walras (1874), one of the founding fathers of mathematical economics.

We refer the reader to the classical analyses of Arrow and Hurwicz (1958); Arrow *et al.* (1959); Negishi (1962) for detailed explanations.

The basic idea behind the *tâtonnement* mechanism is the same assumed in the rudimentary models of sections 1.2 and 1.3, namely that prices of commodities rise and fall in response to discrepancies between demand and supply (the so-called 'law of demand and supply'). In the present case, demand is determined by individual economic agents maximising a utility function subject to a budget constraint, given a certain initial distribution of stocks of commodities. The model can be described schematically as follows.

VARIABLES

$$p(t) : \text{price vector} \quad p \in \mathbb{R}^m_+$$
$$f[p(t)] : \text{vector of excess demand functions}$$

that is, $f_i[p(t)]$, $(i = 1, 2, \ldots, m)$ is the aggregate excess demand for the ith commodity.

ASSUMPTIONS In the continuous-time case, the price adjustment process may be defined by the following system of differential equations

$$\dot{p} = f(p). \tag{3.22}$$

In this simple formulation, we are implicitly assuming that the rate of change of all prices depends on the excess demand according to the same proportionality factor which is set equal to one. This assumption is not essential and will be weakened later.

From an economic point of view, we say that $\bar{p}$ is an 'equilibrium' for the economy if

$$f_i(\bar{p}) \leq 0; \quad \bar{p}_i \geq 0 \text{ for all } i, \quad \bar{p}_j > 0 \text{ for some } j \tag{3.23}$$

and

$$f_j(\bar{p}) < 0 \quad \text{if } \bar{p}_j = 0 \quad (i, j = 1, 2, \ldots, m). \tag{3.24}$$

If the aggregate excess demand functions $f(p)$ are continuous, single-valued (except perhaps at the origin) and all derivatives $\partial f_i/\partial p_j$ exist and are continuous, which we assume here, then system (3.22) possesses solutions and there exists at least one vector $\bar{p}$ satisfying the conditions (3.23) and (3.24).

If we ignore the special ('corner') case (3.24) (i.e., the case in which at price zero there is an excess supply), the equilibrium solution in the economic sense coincides with a fixed point for system (3.22) such that

$$\bar{p}_i \geq 0 \ \forall i \quad \bar{p}_j > 0 \text{ for some } j \quad (i, j = 1, 2, \ldots, m)$$
$$f_i(\bar{p}) = 0 \ \forall i. \tag{3.25}$$

Henceforth, we mean by equilibrium, a set of prices $\bar{p}$ satisfying (3.25).

The economic hypotheses of the model put certain crucial restrictions on the excess demand functions which prove useful in demonstrating existence and stability of a semipositive, equilibrium price vector. First of all, a well-known implication of the hypothesis that agents maximise utility is that the functions $f(p)$ are **homogeneous of degree zero**, namely $f(p) = f(\lambda p)$ for any $\lambda > 0$ (henceforth this will be denoted as hypothesis **H**).

Secondly, consider that the budget constraint for each individual k takes the form

$$\sum_{i=1}^{m} p_i f_i^k(p) = 0$$

where summation is performed over the m commodities and f_i^k denotes the excess demand by the kth economic agent for the ith commodity, i.e., the difference between the agent's demand for, and the agent's initial endowment of, that commodity. Then summing over all economic agents we have

$$\sum_k \sum_i p_i f_i^k(p) = \sum_i p_i f_i(p) = 0$$

which is known as **Walras' Law** (henceforth denoted by **W**). It states that, in view of the budget constraints, for *any* set of semipositive prices p (not necessarily equilibrium prices), the value of aggregate excess demand, evaluated at those prices, must be zero.

LOCAL STABILITY ANALYSIS Local stability of an assumed equilibrium $\bar{p}$, for which $f(\bar{p}) = 0$, can be investigated by means of the linearised system

$$\dot{\xi} = A\xi \tag{3.26}$$

where $\xi \equiv p - \bar{p}$ and

$$A = Df(\bar{p}) = \begin{pmatrix} \frac{\partial f_1(\bar{p})}{\partial p_1} & \frac{\partial f_1(\bar{p})}{\partial p_2} & \cdots & \frac{\partial f_1(\bar{p})}{\partial p_m} \\ \frac{\partial f_2(\bar{p})}{\partial p_1} & \frac{\partial f_2(\bar{p})}{\partial p_2} & \cdots & \frac{\partial f_2(\bar{p})}{\partial p_m} \\ \vdots & \vdots & \vdots & \\ \frac{\partial f_m(\bar{p})}{\partial p_1} & \frac{\partial f_m(\bar{p})}{\partial p_2} & \cdots & \frac{\partial f_m(\bar{p})}{\partial p_m} \end{pmatrix}$$

is the Jacobian matrix of partial derivatives evaluated at the equilibrium. The elements a_{ij} of A denote the effect on the excess demand of the ith commodity owing to a change in the price of the jth commodity.

There exist several known conditions guaranteeing that A is a **stable matrix**, i.e., that all the eigenvalues of matrix A have negative real parts, which, in this case, implies that for (3.26), $\xi = 0$ is asymptotically stable and for system (3.22), $\bar{p}$ is locally asymptotically stable. Some of those conditions are equivalent and some of them can be given an economic interpretation.

Suppose that if the price of the ith commodity increases, while all the other prices remain constant, the excess demand for the ith commodity decreases (and vice versa). Suppose also that the effect of changes in the price of the ith commodity on its own excess demand is stronger than the combined effect of changes in the other prices (where the latter can be positive or negative). This can be formalised by assuming that

$$a_{ii} < 0 \qquad \forall i \tag{3.27}$$

and that there exists a positive vector $d = (d_1, \ldots, d_m)$ such that

$$|a_{ii}|d_i > \sum_{\substack{j=1 \\ j \neq i}}^{m} |a_{ij}|d_j \qquad (i = 1, 2, \ldots, m). \tag{3.28}$$

Assumption (3.28) is often defined as **strict diagonal dominance (SDD)**.[11]

Property **SDD** implies that the symmetric matrix

$$AD + DA^T \tag{3.29}$$

where D is the diagonal matrix whose nonzero entries are the elements d_i, is negative definite. Then, using the Lyapunov function $V(\xi) = \xi^T D \xi$, it can be shown that the matrix A is stable.

Suppose now that we strengthen the assumption on the matrix A by postulating that

$$a_{ij} = \frac{\partial f_i(\bar{p})}{\partial p_j} > 0 \qquad \forall i \neq j \quad (i, j = 1, 2, \ldots, m).$$

Roughly speaking, this means that if we start from equilibrium and the price of a commodity increases (decreases) while the prices of all other

---

[11]The argument that follows could be developed on the alternative assumption that the effect of changes in the price of the ith commodity on the excess demand of that commodity is stronger than its effect on the excess demands of the other commodities, that is: (i) $a_{ii} < 0 \ \forall i$; (ii) there exists a positive vector $d = (d_1, \ldots, d_m)$ such that $|a_{ii}|d_i > \sum_{\substack{i=1 \\ i \neq j}}^{m} |a_{ji}|d_j \ (j = 1, 2, \ldots, m)$.

commodities remain constant, then the excess demand of all of the other commodities increases (decreases). This property, known as **gross substitutability (GS)**, in conjunction with hypothesis **H** implies that $a_{ii} < 0$, and together they imply that A is a Metzler matrix. A matrix A with seminegative diagonal elements and semipositive off-diagonal elements is called a **Metzler matrix** after the American mathematical economist who studied its properties.[12] From this definition, it follows that A is still a Metzler matrix if **GS** is replaced by **weak gross substitutability** $(\partial f_i(\bar{p})/\partial p_j \geq 0$ for all $i \neq j)$.

remark 3.8 Notice that this is a 'local' definition of **GS** appropriate for local stability analysis only. We discuss 'global' **GS** in a moment.

The interested reader will find no less than fifty equivalent necessary and sufficient conditions for stability of Metzler matrices in Berman and Plemmons (1979), pp. 132–64, where proofs of some of the results mentioned in this section are available.[13]

One of these conditions often employed in economic applications is that the leading principal minors of A should alternate in sign, beginning with minus, so that

$$a_{11} < 0, \quad \begin{vmatrix} a_{11} & a_{12} \\ a_{21} & a_{22} \end{vmatrix} > 0, \quad \begin{vmatrix} a_{11} & a_{12} & a_{13} \\ a_{21} & a_{22} & a_{23} \\ a_{31} & a_{32} & a_{33} \end{vmatrix} < 0,$$

$$\ldots, \mathrm{sgn} \begin{vmatrix} a_{11} & a_{12} & \ldots & a_{1m} \\ a_{21} & a_{22} & \ldots & a_{2m} \\ \ldots & \ldots & \ldots & \ldots \\ a_{m1} & a_{m2} & \ldots & a_{mm} \end{vmatrix} = \mathrm{sgn}(-1^m).$$

Finally, notice that Metzler matrices have a property called **D-stability**, i.e., if a Metzler matrix A is stable, so is the matrix DA, where D is any diagonal matrix with positive diagonal elements. It follows that if system (3.22) is locally asymptotically stable, so is the system

$$\dot{p} = \Theta f(p) \tag{3.30}$$

---

[12]Sometimes the term Metzler matrix is used to indicate a matrix whose off-diagonal terms are semi*negative* and the diagonal elements are semi*positive*. Clearly if A is a Metzler matrix according to one definition, $-A$ is Metzler according to the other. Considering that if λ is an eigenvalue of A, then $-\lambda$ is an eigenvalue of $-A$, stability results obtained with one definition can easily be adapted to the alternative definition.

[13]Strictly speaking, in this book we find conditions for positive real parts of *all* eigenvalues of matrices which are **negative Metzler**, i.e., for which signs of coefficients are inverted, cf. n. 12.

where Θ is *any* diagonal matrix with positive elements

$$\Theta = \begin{pmatrix} \theta_1 & 0 & \cdots & 0 \\ 0 & \theta_2 & \cdots & 0 \\ \vdots & \vdots & \ddots & \vdots \\ 0 & 0 & \cdots & \theta_m \end{pmatrix}.$$

The hypothesis of gross substitutability has much deeper implications when applied to global stability analysis of equilibrium. For this purpose **GS** has to be redefined as

$$\frac{\partial f_i(p)}{\partial p_j} > 0 \qquad \forall p \text{ and } \forall i \neq j$$

(which implies $\partial f_i(p)/\partial p_i < 0$). Also we shall consider the slightly modified system of price dynamics (3.30).

First of all, consider that the **GS** hypothesis, together with homogeneity **H** and the Walras Law **W**, imply that there exists a unique, positive equilibrium price for (3.30), up to an arbitrary scale factor. In other words, if we choose one of the commodities as a numéraire, then there exists a unique, positive vector of *relative prices* such that all excess demands are zero.[14] Proof of this statement can be derived from Arrow *et al.* (1959), p. 88, corollary to lemma 1.

GLOBAL STABILITY ANALYSIS Global stability of equilibrium of (3.30) can be studied by means of Lyapunov's second method. For this purpose, choose the distance function

$$V(p) = \sum_{i=1}^{m} \frac{1}{2\theta_i}(p_i - \bar{p}_i)^2. \tag{3.31}$$

Clearly, $V(\bar{p}) = 0$ and $V(p) > 0$ for $p \neq \bar{p}$. Also $V(p) \to \infty$ as $\|p\| \to \infty$ and $V(p)$ is a radially unbounded function.

---

[14] We have a choice here. We can consider **normalised** prices by taking, say, p_1 (assumed to be nonzero) as a numéraire, setting $p_1 = 1$ (which implies $\dot{p}_1 = 0$), and then studying the dynamics of the $(m-1)$ relative prices $\tilde{p}_j$ $(j = 2, \ldots, m)$ where $\tilde{p}_j \equiv p_j/p_1$. In this case, stability refers to the unique, positive equilibrium p. Alternatively, we can consider the m nonnormalised prices p. In this case, if $\bar{p}$ is an equilibrium, any vector $\lambda\bar{p}$ ($\lambda > 0$) will also be an equilibrium. However, the stability properties of all equilibria will be the same so that we can study any of them. With this proviso, we shall take the latter alternative and consider nonnormalised prices.

The derivative of $V(p)$ in (3.31) with respect to time, evaluated along orbits of (3.30) is

$$\dot{V}(p) = \sum_{i=1}^{m} \frac{1}{\theta_i}(p_i - \bar{p}_i)\dot{p}_i$$

$$= \sum_{i=1}^{m} [p_i f_i(p) - \bar{p}_i f_i(p)]$$

$$= -\sum_{i=1}^{m} \bar{p}_i f_i(p) \quad \text{due to property } \mathbf{W}.$$

Thus, a sufficient condition for global stability is

$$\sum_{i=1}^{m} \bar{p}_i f_i(p) > 0 \qquad \forall p \neq \bar{p}.$$

A celebrated result by Arrow *et al.* (1959), pp. 90–3, established that, if $\bar{p} > 0$, and the conditions **GS**, **H** and **W** hold, then for any nonequilibrium $p > 0$ the sufficient condition above is satisfied and therefore $\bar{p}$ is globally asymptotically stable.

remark 3.9 As indicated in Arrow *et al.* (1959), this stability result can be made more general. First of all, the vector of constant speeds of adjustment θ can be replaced by a vector of arbitrary, continuous, sign-preserving functions h. System (3.30) then becomes

$$\dot{p} = h[f(p)]. \tag{3.32}$$

In this case, we must employ a different Lyapunov function that is,

$$V(p) = \max_i \left| \frac{p_i}{\bar{p}_i} - 1 \right|$$

where $\bar{p}_i$ is an element of an equilibrium price vector. This function, known as the **maximum norm** clearly satisfies $V(\bar{p}) = 0$ and $V(p) > 0 \ \forall p \neq \bar{p}$ and is radially unbounded. $\dot{V}$ need not exist however, but, if it does, then we have from (3.32)

$$\dot{V}(p) = -\max_i \left| \frac{h_i[f_i(p)]}{\bar{p}_i} \right|$$

and for $\bar{p} > 0$, $p \neq \bar{p}$, $\dot{V}(p) < 0$ under the hypothesis **H** and a weaker form of substitutability. When the existence of $\dot{V}$ cannot be guaranteed, a more elaborate argument, which we here omit, is required to show convergence to equilibrium.

remark 3.10 Arrow and other authors extended the stability results to the case of weak gross substitutability (see, for example, Arrow and Hurwicz, 1960).

Appendix B: optimal economic growth

Another interesting economic application of the direct method of Lyapunov is the well-known model of equilibrium dynamics in the price and quantity space. There exist a descriptive and an optimal growth version of the model, each of them in a continuous- and a discrete-time version. We shall here consider only the continuous-time optimal growth model, omitting many difficult and intricate technical details and concentrating on the mathematical skeleton of the model and the application of the Lyapunov method that interests us here. The reader who wishes to investigate this problem further can consult, for example, Cass and Shell (1976a, 1976b).

Suppose there is a (benevolent) central planner charged with the responsibility of maximising a certain social utility function over an infinite time horizon. The argument of the utility function is *per capita* consumption and the value of future consumption is discounted at a fixed rate. Output depends on the capital stock (both expressed in *per capita* terms) and the amount of initial capital stock is given. Current output (net of depreciation) can either be consumed or invested to increase the level of capital stock. The planner thus faces a typical economic dilemma: at any instant in time, more output can be allocated to consumption, with a consequent increase in current utility but also a consequent reduction in investment and thus future output. Vice versa, sacrificing current consumption and utility for the sake of greater investment implies higher levels of future production and consumption. Given this broad economic framework, the formal model can be sketched as follows.

VARIABLES

$$k(t) : \text{capital stocks} \quad k \in \mathbb{R}^m_+$$
$$\dot{k}(t) : \text{vector of net capital investment} \quad \dot{k} \in \mathbb{R}^m.$$

ASSUMPTIONS [15]

$\mathcal{U}(k, \dot{k})$: concave utility function

ρ : discount rate of future utility $\rho \in \mathbb{R}_+$

S : convex set of technological restrictions.

THE PROBLEM

$$\max_{k} \int_0^\infty \mathcal{U}(k, \dot{k})e^{-\rho t}dt$$

subject to $(k, \dot{k}) \in S \subset \mathbb{R}^{2m}$ $(P1)$

$$k(0) = k_0$$

where k_0 is an arbitrary initial condition.

Problem $(P1)$ can be attacked by means of the well-known Pontryagin Maximum Principle (Pontryagin *et al.*, 1962). For this purpose, we must first introduce an auxiliary vector-valued variable $q \in \mathbb{R}_+^m$ and define the function

$$H(k, q) = \max_{\dot{k};(k,\dot{k})\in S} [\mathcal{U}(k, \dot{k}) + q\dot{k}]$$ (3.33)

which is called **Hamiltonian** and can be interpreted here as the *current* value of output (consumption plus investment) evaluated in terms of utility (q is then the current price of investment).

According to the Pontryagin Principle, the necessary (though not sufficient) condition for orbits $k(t)$ and $q(t)$ to solve problem $(P1)$ is that they be solutions of the system of differential equations in $\mathbb{R}^{2m}$:

$$\dot{k} = \frac{\partial H(k, q)}{\partial q} = H_q(k, q)$$

                    (3.34)

$$\dot{q} = -\frac{\partial H(k, q)}{\partial k} + \rho q = -H_k(k, q) + \rho q.$$

To 'close' the problem, and select the orbits that solve $(P1)$ among those satisfying (3.34), we add the additional **transversality** condition[16]

$$\lim_{t\to\infty} q(t)k(t)e^{-\rho t} = 0$$ (3.35)

that is, the discounted value of capital must tend to zero asymptotically.

---

[15]$\mathcal{U}$ is an 'indirect' utility function. It depends implicitly on the 'direct' utility function $u(c)$ where c is consumption, on the production function relating output and capital stock and, finally, on the constraint requiring that consumption plus investment cannot exceed output.

[16]The status of the transversality condition in mathematics and economics, and in particular, the question of its sufficiency and necessity for optimality of orbits has been the object of active investigation. We cannot discuss the issue in any detail here and refer the reader to, among others, Chiang (1992) for an introduction.

Suppose now that an isolated stationary solution $(\bar{k}, \bar{q})$ exists, such that

$$\frac{\partial H(\bar{k}, \bar{q})}{\partial q} = 0 \qquad \frac{\partial H(\bar{k}, \bar{q})}{\partial k} = \rho \bar{q}.$$

Given the assumptions on technology and tastes (convexity of the set S and concavity of the utility function $\mathcal{U}$), we do not expect the fixed point $(\bar{k}, \bar{q})$ to be stable in the space (k, q).

To see this, write the Jacobian matrix of system (3.34), evaluated at the fixed point $(\bar{k}, \bar{q})$, in a partitioned form as

$$J = \begin{pmatrix} H_{qk} & H_{qq} \\ -H_{kk} & -H_{qk}^T + \rho I_m \end{pmatrix}$$

where J is a $(2m \times 2m)$ matrix and the $(m \times m)$ submatrices are defined as follows

$$H_{qk} = H_{kq}^T = \frac{\partial H_q}{\partial k}$$

$$H_{qq} = \frac{\partial H_q}{\partial q}$$

$$H_{kk} = \frac{\partial H_k}{\partial k}$$

$$I_m = (m \times m) \text{ identity matrix.}$$

Given the postulated assumptions, the matrices H_{qq} and $-H_{kk}$ are symmetric and positive semidefinite.

The matrix J can be conveniently written as

$$J = J_1 + \frac{\rho}{2} I_{2m}$$

where I_{2m} is the $(2m \times 2m)$ identity matrix and

$$J_1 = \begin{pmatrix} H_{qk} - \frac{\rho}{2} I_m & H_{qq} \\ -H_{kk} & -H_{qk}^T + \frac{\rho}{2} I_m \end{pmatrix}. \tag{3.36}$$

Thus, if λ is an eigenvalue of J, and κ is an eigenvalue of J_1, we have

$$\lambda = \kappa + \frac{\rho}{2}. \tag{3.37}$$

The matrix J_1 has a special structure (which sometimes is also called Hamiltonian) such that if ν_1 is an eigenvalue of J_1, $\nu_2 = -\bar{\nu}_1$ is also an eigenvalue (where $\bar{\nu}_1$ is the complex conjugate of ν_1). From this and (3.37), it follows that it can never be the case that all the eigenvalues of the complete matrix J have negative real parts. Thus $(\bar{k}, \bar{q})$ cannot be asymptotically stable.

However, stability of system (3.34) is not the most interesting question for economic theory. First of all, we are here only concerned with the subset of orbits of (3.34) which are **optimal** in the sense that they solve problem (P1). Secondly, we are mostly interested in the behaviour of the real variables (the capital stock) rather than that of prices.

The Lyapunov method is very useful to prove that, under certain conditions, the stationary capital stock $\bar{k}$ is asymptotically stable when the dynamics of (3.34) is constrained to an appropriately reduced m-dimensional space.

To construct the reduced system, we define a **value function**

$$W(k_0) = \int_0^\infty \mathcal{U}(k^*(t, k_0), \dot{k}^*(t, k_0)) e^{-\rho t} dt$$

with $k^*(t, k_0)$ an orbit starting at k_0 and solving (3.34) subject to (3.35). In what follows, we omit the reference to the initial value k_0 and denote by $k^*(t)$, $q^*(t)$ the orbits satisfying (3.34) and (3.35) (the **optimal paths** or **orbits**). We also indicate dependence on time of k^* and q^* only when necessary. Then, assuming that $W(k) \in C^2$, along an optimal path,[17]

$$q^*(t) = W'[k^*(t)]$$

and therefore

$$\dot{q}^*(t) = W''[k^*(t)]\dot{k}^*(t).$$

Thus, the m-dimensional system

$$\dot{k}^* = H_q[k^*, W'(k^*)] \tag{3.38}$$

can be used to study the dynamical behaviour of the optimal orbits and, in particular, the stability properties of the stationary orbit $k^*(t) = \bar{k}$. First, we notice that $H_q[\bar{k}, W'(\bar{k})] = H_q(\bar{k}, \bar{q}) = 0$ and $\bar{k}$ is a fixed point of (3.38).

Its stability can be investigated by using the Lyapunov function

$$V(k^*) = -(k^* - \bar{k})(q^* - \bar{q}) = -(k^* - \bar{k})(W'(k^*) - W'(\bar{k})).$$

Clearly $V(\bar{k}) = 0$. Also, if $\mathcal{U}$ is concave, so is W and therefore

$$V(k^*) \geq 0 \quad \text{for} \quad k^* \neq \bar{k},$$

strict inequality holding for strictly concave functions. Then, if we prove that $\dot{V} < 0$ along orbits of (3.38), we can conclude that the stationary solution $\bar{k}$ of (3.38) is asymptotically stable. Sufficient conditions for $\dot{V} < 0$ (thus, for asymptotic stability of $\bar{k}$) are discussed in Cass and Shell (1976a), pp. 55–60, where the authors point out that optimal paths, in general, need not converge to a stationary point.

[17]This result was proven by Arrow and Kurz (1970). For a discussion of the conditions under which $W(k)$ is differentiable, see Benveniste and Scheinkman (1979).

To illustrate the point, we discuss here the simpler, but interesting, special case in which the differential equations in (3.34) are linear. This happens when the Hamiltonian function is quadratic in the variables (k, q). (Alternatively, we can think of those equations as a linear approximation around an assumed fixed point $(\bar{k}, \bar{q})$. In this case, the variables must be re-interpreted as discrepancies from equilibrium values and the results of the analysis hold only locally.) In the linear case the differential equations are

$$\begin{pmatrix} \dot{k} \\ \dot{q} \end{pmatrix} = \begin{pmatrix} A & B \\ C & -A^T + \rho I_m \end{pmatrix} \begin{pmatrix} k \\ q \end{pmatrix} \tag{3.39}$$

where A, B, C are the constant $(m \times m)$ matrices

$$A = H_{qk}$$
$$B = H_{qq}$$
$$C = -H_{kk}.$$

System (3.39) has a unique fixed point at $(0, 0)$. Optimal paths must satisfy (3.39) and also the equation

$$q = Xk \tag{3.40}$$

where $X = W''(\bar{k})$ is a constant $(m \times m)$ matrix. (To simplify notation, henceforth, we omit the asterisk in referring to the optimal path.) Given the hypotheses on tastes (utility function) and technology (production function), the matrices $B, C, -X$ are symmetric, positive semidefinite but, for simplicity's sake, we assume here that they are positive definite.

From (3.39) and (3.40) we have

$$\dot{q} = X\dot{k} = XAk + XBq = (XA + XBX)k$$
$$= Ck + (\rho I_m - A^T)q = (C + \rho X - A^T X)k.$$

Hence, along optimal paths we must have

$$P(X)k = (XA + XBX - C + A^T X - \rho X)k = 0 \tag{3.41}$$

and, for $k \neq 0$, (3.41) is satisfied if X solves the matrix equation $P(X) = 0$.[18]

Let us now perform the following change of variables

$$\hat{q} = q - Xk \tag{3.42}$$

that is, $\hat{q}$ denotes the difference between q and its optimal value defined by

---

[18]The matrix equation $P(X) = (XA + XBX - C + A^T X - \rho X) = 0$ is known in the literature on linear-quadratic dynamic optimisation as the **Riccati equation**. See, for example, Wimmer (1984).

(3.40). It follows that

$$\begin{pmatrix} \dot{k} \\ \dot{q} \end{pmatrix} = \begin{pmatrix} I_m & 0 \\ -X & I_m \end{pmatrix} \begin{pmatrix} \dot{k} \\ \dot{q} \end{pmatrix}. \tag{3.43}$$

Pre-multiplying both sides of (3.39) by the matrix

$$\begin{pmatrix} I_m & 0 \\ -X & I_m \end{pmatrix}$$

and using (3.41), (3.42) and (3.43), we obtain

$$\begin{pmatrix} \dot{k} \\ \dot{\hat{q}} \end{pmatrix} = \begin{pmatrix} A + BX & B \\ 0 & -(A + BX)^T + \rho I_m \end{pmatrix} \begin{pmatrix} k \\ \hat{q} \end{pmatrix}.$$

Along optimal paths where (3.40) is satisfied, $\hat{q} = 0$ and therefore $\dot{\hat{q}} = 0$. Hence convergence of optimal orbits to the unique fixed point can be studied by means of the reduced linear system

$$\dot{k} = (A + BX)k. \tag{3.44}$$

Stability of (3.44) obviously depends on the properties of the matrix $A+BX$. We list below a few examples of sufficient conditions for stability.

(i) $\rho = 0$. In this case, for $k \neq 0$ we have

$$P(X) = XBX + A^T X + XA - C = 0$$

whence

$$XBX + [XA + A^T X] = C.$$

Adding XBX on both sides and factoring we obtain

$$X[A + BX] + [A^T + XB]X = C + XBX.$$

Consider now that X is negative definite, XBX is positive definite if B is, and therefore $C + XBX$ is positive definite. We can now choose the Lyapunov function $V(k) = -k^T X k$ and conclude that $A + BX$ is a stable matrix (i.e., all its eigenvalues have negative real parts).

(ii) The matrix $B^{-1}A$ is negative definite. In this case (3.44) can be written as

$$\dot{k} = B[B^{-1}A + X]k = Mk.$$

Then M is the product of the positive-definite matrix B and the negative definite matrix $(B^{-1}A + X)$ and therefore, M is a stable matrix, independently of ρ.

(iii) If $\rho > 0$, from (3.41) we can write

$$G \equiv X[A + BX] + [A^T + XB]X = C + XBX + \rho X.$$

Therefore, if $C + XBX + \rho X$ is positive definite, stability of (3.44) can be again proved by choosing the Lyapunov function $-k^T Xk$. To study this condition in greater detail consider that for an arbitrary, nonzero real vector $x \in \mathbb{R}^m$ and a real, symmetric $(m \times m)$ matrix A we have

$$x^T Ax \geq \lambda_m(A)$$

where $\lambda_m(A)$ is the smallest (real) eigenvalue of A. Then

$$x^T Gx = x^T(C + XBX + \rho X)x \geq \lambda_m(C) + \lambda_m(B) + \rho\lambda_m(X).$$

Hence, if

$$\rho < \frac{\lambda_m(B) + \lambda_m(C)}{-\lambda_m(X)}$$

$x^T Gx > 0$ and consequently, choosing again $V(x) = -k^T Xk$ as a Lyapunov function, we can establish that $(A + BX)$ is a stable matrix.

Appendix C: manifolds and tangent spaces

In the study of dynamical systems we often discuss the action of maps or flows on spaces called manifolds. Although we cannot discuss the issue in detail here, a rudimentary explanation of the basic notions is necessary. Roughly speaking, a **manifold** is a set which can be made an Euclidean space by means of local coordinates. A more formal definition is the following.

definition 3.9 *A C^r m-dimensional **manifold** M is a metric space together with a system of homeomorphisms ϕ_i such that: (i) each ϕ_i is defined on an open subset $W_i \subset M$ and maps it to an open subset $\phi_i(W_i) \subset \mathbb{R}^m$; (ii) $M = \cup_i W_i$; (iii) if the intersection $W_i \cap W_j \neq \emptyset$, then the map $\phi_i \circ \phi_j^{-1} : \phi_j(W_i \cap W_j) \subset \mathbb{R}^m \rightarrow \phi_i(W_i \cap W_j) \subset \mathbb{R}^m$ is C^r. A map ϕ_i is called a **coordinate chart**.*

Two important related concepts that we use in the following pages are those of tangent space and tangent vector. Consider a manifold M and an associated coordinate chart ϕ_i. A map $\gamma : [-\delta, \delta] \subset \mathbb{R} \rightarrow M$ defines an arc or curve on M with endpoints $\gamma(-\delta)$ and $\gamma(\delta)$. The curve γ is of class C^k

if, for any coordinate chart ϕ_i, the function $\phi_i \circ \gamma$ is C^k. If $\gamma(0) = p \in M$, we say that

$$v_p^i = (\phi_i \circ \gamma)'(0)$$

is the **tangent vector** to γ at p. (Different vectors generated by a curve in different coordinate charts form an equivalence class and they are identified.)

The set $T_pM = \{v_p|v_p$ is a tangent vector to a differentiable curve at $p \in M\}$ is called the **tangent space** at p.

For each $p \in M$, T_pM is a vector space endowed with the norm $\|v_p\| = (v_p, v_p)^{1/2}$, where as usual, $(\cdot, \cdot)$ denotes scalar product.

Exercises

3.1   Consider a saddle point of a linear system of differential equations in $\mathbb{R}^2$. Explain why it is unstable.

3.2   For the following differential equations: sketch the function in the $(x, \dot{x})$ plane; mark the fixed points (where $\dot{x} = 0$); place arrows on the abscissa indicating whether x is converging to or diverging from the equilibria; label each fixed point as stable or unstable

(a) $\dot{x} = x^3$
(b) $\dot{x} = 0.5(x^2 - 1)$
(c) $\dot{x} = (x - 1)(x - 3)$
(d) $\dot{x} = 4x$
(e) $\dot{x} = 2x(1 - x)$
(f) $\dot{x} = \frac{1}{1-x^2}$.

3.3   For the following difference equations: sketch the function in the (x_n, x_{n+1}) plane; draw in the bisector line (where $x_{n+1} = x_n$) and mark the fixed points; choose a few initial values and iterate a few times and use arrows to indicate convergence to or divergence from the equilibria; label each fixed point as stable or unstable. Notice that although these difference equations are similar to some of the differential equations in problem 1, the fixed point stability is not necessarily the same

(a) $x_{n+1} = x_n^3$
(b) $x_{n+1} = 4x_n$
(c) $x_{n+1} = 2x_n(1 - x_n)$.

3.4   Find the fixed points of the following nonlinear differential systems. Use the linear approximation and the Hartman–Grobman theorem to describe the local stability of each system at each fixed point

(a)
$$\dot{x} = y^2 - 3x + 2$$
$$\dot{y} = x^2 - y^2$$

(b)
$$\dot{x} = -y$$
$$\dot{y} = x - x^5$$

(c)
$$\dot{x} = x + x^2 + xy^2$$
$$\dot{y} = y + y^{3/2}$$

(d)
$$\dot{x} = y - x^2 + 2$$
$$\dot{y} = 2(x^2 - y^2)$$

(e)
$$\dot{x} = -y + x + xy$$
$$\dot{y} = x - y - y^2$$

(f)
$$\dot{x} = y$$
$$\dot{y} = -(1 + x^2 + x^4)y - x$$

(g)
$$\dot{x} = \alpha x - \beta xy$$
$$\dot{y} = -\gamma y + \delta xy \qquad \alpha, \beta, \gamma, \delta \in \mathbb{R}_+.$$

3.5   Consider the second-order nonlinear differential equations in $x(t)$ below. Determine the fixed points of each of them, indicate the dimensions of the stable, unstable and centre manifolds

(a)
$$\ddot{x} + x - x^3 = 0.$$

(b)
$$\ddot{x} - \dot{x} + x^2 - 2x = 0.$$

3.6   Consider the following discrete-time, first-order systems in the variables (x_n, y_n, z_n). Determine the fixed points and for each indicate
the dimensions of the stable, unstable and centre manifolds and the
relative position of the eigenvalues in the complex plane. Describe
the local stability for each fixed point.

(a)

$$x_{n+1} = x_n^2 - x_n y_n$$
$$y_{n+1} = y_n^{1/2} + 6$$
$$z_{n+1} = z_n^2 + x_n y_n$$

(b)

$$x_{n+1} = 0.5 x_n^2$$
$$y_{n+1} = -y_n + 3$$
$$z_{n+1} = z_n - 2$$

(c)

$$x_{n+1} = x_n y_n + 30$$
$$y_{n+1} = 2 y_n^{1/2} + 8$$
$$4 z_{n+1} = x_n y_n + 4 z_n^2 + 8.$$

3.7   It will have been observed that in exercise 3.4(g) the more interesting
fixed point in the positive quadrant is nonhyperbolic and, therefore,
the Hartman–Grobman theorem is not applicable. However, that
system is one of the few nonlinear systems for which solution curves
in the (x, y) plane can be determined by integrating the differential
equation $\frac{dy}{dx} = \frac{\dot{y}}{\dot{x}}$. Find the exact solutions to the following systems
for which $\frac{\dot{y}}{\dot{x}}$ can be easily integrated

(a)

$$\dot{x} = \alpha x - \beta x y$$
$$\dot{y} = -\gamma y + \delta x y \qquad \alpha, \beta, \gamma, \delta \in \mathbb{R}_+$$

(b)

$$\dot{x} = \sin x$$
$$\dot{y} = \cos x \sin x$$

(c)

$$\dot{x} = x y$$
$$\dot{y} = 4(y - 1).$$

3.8   Find the expression for the nontrivial fixed points of the *logistic map*

$$x_{n+1} = \mu x_n(1 - x_n)$$

and give parameter values over which the fixed points are locally stable using the linear approximation. Sketch the fixed point curves in the plane $(\bar{x}, \mu)$ over the domain $\mu \in [-1, 3]$.

3.9   Consider the system

$$\begin{pmatrix} \dot{x} \\ \dot{y} \end{pmatrix} = \begin{pmatrix} 0 & 1 \\ 0 & 0 \end{pmatrix} \begin{pmatrix} x \\ y \end{pmatrix}.$$

Find the fixed points and discuss their stability.

3.10   For the following systems: show that $V(x, y) = \frac{1}{2}(x^2 + y^2)$ is a Lyapunov function for the equilibrium at $(0, 0)$; give the domain of V for which the equilibrium is asymptotically stable

(a)

$$\dot{x} = x - 2y^2$$
$$\dot{y} = 2xy - y^3$$

(b)

$$\dot{x} = -x + y^2$$
$$\dot{y} = -xy - x^2$$

(c)

$$\dot{x} = -y - x^3$$
$$\dot{y} = x - y^3$$

(d)

$$\dot{x} = -x(x^2 + y^2 - 2)$$
$$\dot{y} = -y(x^2 + y^2 - \frac{2x^2}{y^2} + 3)$$

(e)

$$\ddot{x} + \dot{x} - \frac{\dot{x}^3}{4} + x = 0.$$

3.11   Find a Lyapunov function for each of the following systems and indicate the domain for which the equilibrium at $(0, 0)$ is asymptotically stable

(a)

$$\dot{x} = x^2 - y$$
$$\dot{y} = 3x^3$$

(b)

$$\dot{x} = -x - y^2$$
$$\dot{y} = kxy$$

(c)

$$\dot{x} = y$$
$$\dot{y} = -y - x^3$$

(d)

$$\dot{x} = y + kx(x^2 + y^2)$$
$$\dot{y} = -x.$$

3.12 Prove that the origin is an asymptotically stable fixed point for the following system using the rule of thumb discussed in section 3.3 (from Glendinning, 1994, p. 52)

$$\dot{x} = -y - x^3$$
$$\dot{y} = x^5.$$

3.13 Prove (3.18) and (3.19).

3.14 From Appendix A, prove that (3.29) is negative definite and therefore, that **SDD** implies that A is a stable matrix.

3.15 From Appendix B, prove that the matrix J_1 defined in (3.36) has a special structure such that if ν_1 is an eigenvalue of J_1, $\nu_2 = -\bar{\nu}_1$ is also an eigenvalue (where if ν is complex, $\bar{\nu}$ is its conjugate).

4

Invariant and attracting sets, periodic and quasiperiodic orbits

In chapter 3 we discussed the behaviour of a dynamical system when it is displaced from its state of rest, or equilibrium, and, in particular, we studied the conditions under which the displaced system does not wander too far from equilibrium or even converges back to it as time goes by. For such cases, we call the equilibrium stable or asymptotically stable. But what happens if we perturb an unstable equilibrium?

For an autonomous linear system, if we exclude unlikely borderline cases such as centres, the answer to this question is straightforward: orbits will diverge without bound.

The situation is much more complicated and interesting for nonlinear systems. First of all, in this case we cannot speak of *the* equilibrium, unless we have established its uniqueness. Secondly, for nonlinear systems, stability is not necessarily global and if perturbations take the system outside the basin of attraction of a locally stable equilibrium, it will not converge back to it. Thirdly, besides convergence to a point and divergence to infinity, the asymptotic behaviour of nonlinear systems includes a wealth of possibilities of various degrees of complexity.

As we mentioned in chapter 1, closed-form solutions of nonlinear dynamical systems are generally not available, and consequently, exact analytical results are, and will presumably remain, severely limited. If we want to study interesting dynamical problems described by nonlinear differential or difference equations, we must change our orientation and adapt our goals to the available means. The short-run dynamics of individual orbits can often be described with sufficient accuracy by means of straightforward numerical integration of the differential equations, or iteration of the maps. However, in applications, we are usually concerned not with short-term properties of individual solutions, but with the long-run global, qualitative properties of bundles of solutions which start from certain practically relevant subsets of initial conditions. Those properties can be investigated effectively by studying the asymptotic behaviour of orbits and concentrating the analysis on regions of the state space which are *persistent* in the weak sense that orbits never leave them, or in the stronger sense that orbits are attracted to them.

The asymptotic regime of a system is the only observable behaviour, in the sense that it is not ephemeral, can be repeated and therefore be *seen* (e.g., on the screen of a computer). The structure of the limit set of a dynamical system is often easier to investigate than its overall orbit structure. For most practical problems, moreover, an understanding of limit sets and their basins of attraction is quite sufficient to answer all the relevant questions.

Even though transient behaviour may sometimes last for a very long time and be an interesting topic of investigation, in the following chapters we deal with the subject only occasionally, and concentrate on the long-term behaviour of dynamical systems.

4.1 Invariant and limit sets

In order to discuss the asymptotic behaviour of dynamical systems, we need to develop some basic concepts which are common to both flows and maps and which describe, in a sense, different kinds or degree of persistence of subsets of the state space. The definitions that follow are given, generally, for flows on metric spaces, and we point out the adaptations for discrete-time maps, whenever necessary. We begin by defining invariant sets in a more rigorous manner than that used in previous chapters.

definition 4.1 *Given a flow ϕ on a metric space M, we say that $S \subset M$ is* **invariant** *for ϕ if, for any $x \in S$, we have $\phi(t, x) \in S$ for all $t \in \mathbb{R}$. The set S is called* **positively** *(or* **forward**) **invariant** *if the property is restricted to $t \in \mathbb{R}_+$ and* **negatively** *(or* **backward**) **invariant** *if $t \in \mathbb{R}_-$.*

Essentially the same definition applies to discrete-time systems defined by a map G with the proviso that t is replaced by $n \in \mathbb{Z}$ ($n \in \mathbb{Z}_+$ or $n \in \mathbb{Z}_-$).

Thus, an invariant set is persistent in the sense that an orbit starting from a point in the set will never leave it, *if there are no exogenous disturbances*.

Certain invariant sets play an important role in the organisation of the orbit structure in the state space. In chapter 2 we saw that, generically, when the unique equilibrium point of a linear system of differential equations is asymptotically stable, all orbits starting off equilibrium converge to it for $t \to \infty$, and they will diverge to $\pm\infty$ as $t \to -\infty$. Figuratively speaking, the equilibrium is the typical orbit's *long-run future*, whereas (plus or minus) infinity is its *remote past*. The reverse is true for linear systems characterised by a **totally unstable** equilibrium, that is, systems whose controlling matrix has *all* eigenvalues with positive real parts. Instead, for linear systems with a saddle point, excepting special sets of initial conditions, plus or minus infinity is both the long-run future and the remote past of the system. Analogous considerations apply to linear systems of difference equations.

For nonlinear systems, in general the structure of the *long-run future* and *remote past* of orbits is more complex and requires the introduction of more precise concepts. The definitions given below for a flow ϕ are essentially the same as for a map G. Unless we indicate otherwise, we assume that G is a smooth, invertible map (diffeomorphism) and mention, from time to time, the special problems arising for noninvertible maps.

definition 4.2 *Let $\phi(t,x)$ be a flow on a metric space M. Then a point $y \in M$ is called a $\boldsymbol{\omega}$-limit point of $x \in M$ for $\phi(t,x)$ if there exists a sequence $\{t_i\}$ increasing in i to infinity, such that*

$$\lim_{i \to \infty} d(\phi(t_i, x), y) = 0$$

where d is a distance function on M. The set of all ω-limit points of x for $\phi(t,x)$ is called the $\boldsymbol{\omega}$-limit set and denoted by $\omega(x)$.

The definitions of $\boldsymbol{\alpha}$-limit point and the $\boldsymbol{\alpha}$-limit set of a point $x \in M$ are derived from definition 4.2 trivially by taking sequences $\{t_i\}$ decreasing in i to minus infinity. The α-limit set of x is denoted as $\alpha(x)$. For example, in the special case of a linear system of differential equations defined on $\mathbb{R}^m$ and characterised by an asymptotically stable fixed point $\bar{x}$, the ω-limit point and the ω-limit set of all $x \in \mathbb{R}^m$ are simply $\bar{x}$. Were $\bar{x}$ totally unstable, it would be the α-limit point and α-limit set of all $x \in \mathbb{R}^m$.

Analogously, we can define ω-limit sets for sets larger than a point, for example, an asymptotically stable limit cycle Γ (on which more in section 4.4). In this case, for each point $x_j \in \Gamma$, there are time sequences $\{t_1^{(j)}, t_2^{(j)}, \ldots, t_k^{(j)}, \ldots\}$ such that, starting from a point x_0 in a neighbourhood of Γ (its basin of attraction), the orbits $\gamma_j = \{\phi(t_i^{(j)}, x_0) | i = 1, \ldots, k\}$ converge to x_j as $k \to \infty$. The set of all these points x_j, i.e., the set Γ, is the ω-limit set of x_0. An example of an ω-limit set which is a limit cycle is illustrated in figure 4.1. The definitions of α-limit and ω-limit points and sets for diffeomorphisms are essentially the same with the obvious difference that now the sequences $\{t_i\}$, $t_i \in \mathbb{R}$ are replaced by $\{n_i\}$, $n_i \in \mathbb{Z}$.

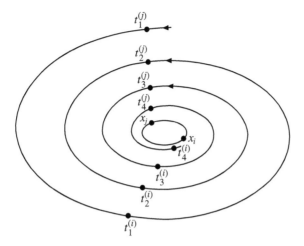

Fig. 4.1 The ω-limit set Γ, a limit cycle

remark 4.1 Notice that a **forward orbit** starting from a point $x \in M$ (that is, a set $\{\phi(t, x) \mid t \geq 0\}$ or $\{G^n(x) \mid n \geq 0\}$) may or may not have points in common with the ω-limit set $\omega(x)$. For example, consider the orbit generated by a map G, starting from a point x_0 and asymptotically converging to a fixed point $\bar{x}$. Then $\omega(x_0) \cap \{G^n(x_0) \mid n \geq 0\} = \emptyset$. On the other hand, if $x_0 \in \omega(x_0)$, then a forward orbit starting from x_0 is contained in its limit set and may even coincide with it. For example, this occurs when x_0 is itself a fixed or a periodic point of G. In fact, if G is invertible and x_0 is a periodic point of period $n \geq 1$, $\omega(x_0) = \alpha(x_0) = \{G^k(x_0) \mid 0 \leq k \leq n - 1\}$. Points $x \in M$ such that $x \in \omega(x)$ are sometimes called **positively recurrent**; they are called **negatively recurrent** if $x \in \alpha(x)$ and simply **recurrent** if they are both positively and negatively recurrent.

For a noninvertible map G for which the inverse G^{-1} is not well defined, to construct the α-limit points and sets of a point x, we must make a choice of pre-images of x (see the definition of the unstable manifold of a noninvertible map, chapter 5, section 5.2, equation (5.7)). Some of the consequent problems are occasionally discussed under the label 'backward dynamics', a situation often encountered in economic models represented by noninvertible maps (cf. for example, Medio, 2001).

There exist some interesting general properties of α- and ω-limit sets. We list them for a flow ϕ on a metric space M.

(i) If the forward orbit of ϕ, $\{\phi(t, x) \mid t \geq 0\}$ is contained in a compact[1] subset of $A \subset M$, then $\omega(x)$ is nonempty, compact and connected.[2] Accordingly, if the backward orbit $\{\phi(t, x) \mid t \leq 0\}$ is contained in a compact set, $\alpha(x)$ is nonempty, compact and connected.

(ii) Under the same conditions as in (i),

$$\lim_{t \to \infty} d[\phi(t, x), \omega(x)] = 0$$

$$\lim_{t \to -\infty} d[\phi(t, x), \alpha(x)] = 0.$$

(iii) If $x \in S$, where S is a closed and positively invariant subset of the state space M, then $\omega(x) \subset S$. Analogously, if $x \in S$, where S is closed and negatively invariant, then $\alpha(x) \subset S$.

(iv) If a point $y \in M$ belongs to the ω-limit set of another point $x \in M$, then the α- and ω-limit set of the former must be contained, respectively, in the α- and ω-limit set of the latter, i.e., $\alpha(y) \subset \alpha(x)$ and $\omega(y) \subset \omega(x)$.

The same properties hold for a diffeomorphism G on M, with the exception that α- or ω-limit sets for maps need not be connected. If G is not invertible, read properties (i)–(iv) only for the parts concerning the ω-limit sets.

If we consider the ensemble of all α- and ω-limit sets of points $x \in M$ under the action of a flow ϕ, we can define the **positive limit set** of ϕ as

$$L^+(\phi) = \mathrm{cl}(\cup_{x \in M} \omega(x))$$

(where $\mathrm{cl}(A)$ denotes the **closure** of the set A, i.e., the smallest closed set containing A) and the **negative limit set** of ϕ as

$$L^-(\phi) = \mathrm{cl}(\cup_{x \in M} \alpha(x)).$$

The sets $L^+(\phi)$ and $L^-(\phi)$ are ϕ-invariant and describe the asymptotic behaviour of the orbits generated by ϕ in the sense that, for every $x \in M$, $\phi(t, x)$ converges to $L^+(L^-)$ as $t \to +\infty(-\infty)$. Analogous definitions apply to invertible maps.

---

[1] A rigorous definition of compactness requires first a definition of cover. If X is a metric space, we say that a collection $\{V_i\}_{i \in I}$ (I any set) of open subsets of X is an **open cover** for $K \subset X$ if $K \subset \cup_{i \in I} V_i$. If I is finite, $\{V_i\}_{i \in I}$ is called a **finite open cover**. The set K is **compact** if (and only if) every open cover has a finite subcover, namely $K \subset \cup_{i \in I} V_i \Rightarrow \exists J \subset I$, J finite such that $K \subset \cup_{i \in J} V_i$. A well-known result in topology states that subsets of $\mathbb{R}^m$ are compact if and only if they are *closed* and *bounded*. See, for example, Sutherland (1999), pp. 82–6.

[2] We say that a subset S of a metric space M is **connected** if there are *no* nonempty sets A and B such that $S = A \cup B$ and there exists a $k > 0$ such that $d(A, B) > k$. Recall that $d(A, B)$ here means $\min_{x \in A, y \in B} d(x, y)$. From this it follows that an orbit for a flow is connected, but an orbit for a map need not be. For example, a limit cycle of a system of differential equations is connected whereas a periodic orbit of a system of difference equations is not.

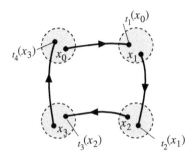

Fig. 4.2 A chain recurrent set with $k = 3$

We now present two related concepts capturing interesting recurrence properties of orbits, which are weaker than those of limit sets.

definition 4.3 *A point $x \in M$ is called* **nonwandering** *if, for any open neighbourhood N of x:*

(a) *(flows) there exists a $t \neq 0$ such that $\phi(t, N) \cap N \neq \emptyset$; or*
(b) *(maps) there exists an integer $n \neq 0$ such that $G^n(N) \cap N \neq \emptyset$. (If G is noninvertible we only take $n > 0$.)*

The union of all nonwandering points is called a **nonwandering set** *and is denoted by $\Omega(\phi)$. The set $\Omega(\phi)$ is closed and ϕ-invariant.*

Weaker yet is the notion of chain recurrence (see figure 4.2).

definition 4.4 *A point $x \in M$ is called* **chain recurrent** *for a flow ϕ if, for every $\epsilon > 0$, there exists a sequence $\{x_0, x_1, x_2, \ldots, x_k = x_0; t_1, t_2, \ldots, t_k\}$ with $t_j \geq 1$, such that for $1 \leq j \leq k$, $d[\phi(t_j, x_{j-1}), x_j] < \epsilon$. The* **chain recurrent set** *for ϕ is the set*

$$R(\phi) = \{x \in M \mid x \text{ is chain recurrent}\}.$$

Entirely analogous definitions can be given for maps.

From the definitions above it follows that for a flow or a map, the chain recurrent set contains the nonwandering set which, in its turn contains the union of the positive and negative limit sets, that is, $L^-(\phi) \cup L^+(\phi) \subset \Omega(\phi) \subset R(\phi)$.

Consider the example illustrated in figure 4.3. In polar coordinates (r, θ) we have the system

$$\dot{r} = r(1 - r^2) \tag{4.1}$$
$$\dot{\theta} = \theta^2 (2\pi - \theta)^2. \tag{4.2}$$

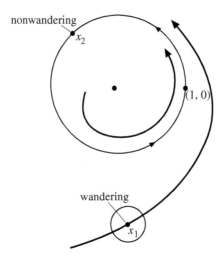

Fig. 4.3 Wandering and nonwandering points

Because r is essentially nonnegative, (4.1) implies that $\dot{r} \gtrless 0$ if $1 \lessgtr r$. Thus, in an open neighbourhood V of the closed curve Γ, orbits are attracted to it. A point x_1 on the outer orbit is wandering because for sufficiently small neighbourhoods of x_1, $N(x_1)$, there exists $t_0 > 0$ such that $N(x_1) \cap \phi[t, N(x_1)] = \emptyset$ for all $t > t_0$, where ϕ denotes the flow generated by (4.1)–(4.2). On the contrary, points on the curve Γ are nonwandering. For all points $x \in V$ the ω-limit set $\omega(x)$ is $\{(1,0)\}$, which is contained in the nonwandering set $\Omega(\phi) = \Gamma$.

Not all parts of the limit sets are equally interesting. We are here particularly concerned with subsets of the positive limit sets which are attracting. The notion of attractiveness is intimately related to those of stability and observability. There are various, not entirely equivalent notions of attractiveness and we consider a few basic ones in the next section.

4.2 Stability and attractiveness of general sets

The first group of definitions of stability and attractiveness are based on an extension of the concepts of stability or asymptotic stability to *general sets*, i.e., sets which are not necessarily single points. There exist several, more or less equivalent formulations of those concepts and there are many aspects and details that could be discussed. Here we focus our attention on those properties most relevant to the topics presented in this and the succeeding chapters.

definition 4.5 *A compact subset A of a metric space M, positively invariant with respect to the flow ϕ is said to be* **stable** *if, for every neighbourhood U of A, there exists a neighbourhood V such that any forward orbit starting in V is entirely contained in U, that is to say, we have*

$$\phi(t, V) \subset U \qquad t \geq 0.$$

definition 4.6 *The set A of definition 4.5 is said to be* **asymptotically stable** *if*

(a) *it is stable and moreover,*
(b) *it is* **attracting** *for the flow ϕ in the sense that the set*

$$B(A) = \{x \in M| \lim_{t \to \infty} d\,[\phi(t, x), A] = 0\}$$

is a neighbourhood of A.

Condition *(b)* of definition 4.6 (attractiveness) can be replaced by one or the other of the following:[3]

(b′) *the set $B(A) = \{x \in M| \omega(x) \subset A\}$ is a neighbourhood of A; or*
(b″) *the set $B(A) = \{x \in M|$ for every neighbourhood U of A there exists a neighbourhood N of x and $T > 0$ such that $\phi(t, N) \subset U$ for $t \geq T\}$ is a neighbourhood of A; or*
(b‴) *the set A is attracting for a flow ϕ if there exists a* **trapping region** *$U \subset M$ (that is, a positively invariant set such that $\phi[t, \mathrm{cl}(U)] \subset \mathrm{int}(U)$ for $t > 0$), such that*

$$A = \bigcap_{t \geq 0} \phi(t, U).$$

Notice that if stability (or asymptotic stability) is *uniform* (as is always the case for autonomous systems of differential equations), then the choice of the neighbourhood V in definition 4.5 and the time T in definition 4.6 *(b″)* are independent of the initial time t_0. Analogously, convergence defined in definition 4.6 *(b)* is uniform in t_0.

The set $B(A)$ is called the **basin of attraction**, **stable set**, or even **stable manifold** of A, although it need not be a manifold in a strict sense (see definition 3.9 in appendix C of chapter 3). Entirely analogous definitions of stability and asymptotic stability can be written for (discrete-time) maps with only slight changes.

---

[3] The symbol $\mathrm{int}(A)$ indicates the **interior** of a subset $A \subset \mathbb{R}^m$, that is

$$\mathrm{int}(A) = \{x| \exists \epsilon > 0 \text{ such that } N(\epsilon, x) \text{ contains only points in } A\}$$

where $N(\epsilon, x)$ is the set of points whose distance from x is less than ϵ. That is, $\mathrm{int}(A)$ is the set of points in A which have neighbourhoods entirely contained in A.

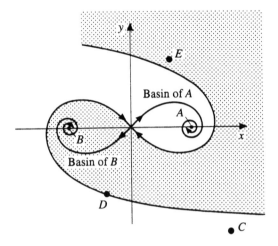

Fig. 4.4 Basins of attraction for fixed points A and B

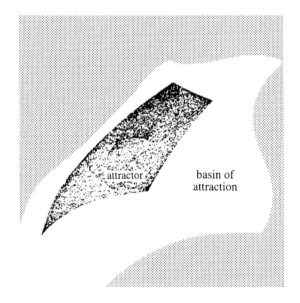

Fig. 4.5 A chaotic attractor and its basin of attraction. Reproduced by
permission of Cambridge University Press

An example of stable, attracting fixed points and their basins of attraction
for a flow in $\mathbb{R}^2$ is provided in figure 4.4 where points C, D and E are referred
to in exercise 4.1. A more complicated attracting set for a two-dimensional
map, and its basin of attraction are depicted in figure 4.5.

In the definitions above, an asymptotically stable set A is characterised by the same two basic properties of an asymptotically stable point, namely:

(i) *orbits starting near A stay near A;*
(ii) *a whole neighbourhood of A is attracted to it.*

remark 4.2 Notice that, when the invariant set A is not a single point, stability (or asymptotic stability) of A does *not* necessarily imply the same properties for *individual points* of A. This can be seen by considering a simple example of a two-dimensional system of differential equations characterised by an equilibrium (fixed) point $\bar{x}$ of the *centre* type. We know that around $\bar{x}$ there exists a continuum of invariant periodic orbits. Let Γ be any one of them with period τ. According to our hypothesis, the set $\Gamma = \{\phi(t, x) | x \in \Gamma, t \in [0, \tau)\}$ is stable (though not asymptotically stable). That is, any orbit starting sufficiently near Γ will remain arbitrarily close to it. But let us now choose a point $x_0 \in \Gamma$ and a second point $y_0 \notin \Gamma$ but close to x_0. In general, their distance cannot be kept within an arbitrarily chosen value, that is, for an arbitrarily given $\epsilon > 0$ there may not be any $\delta(\epsilon)$ such that, if $d(x_0, y_0) < \delta$, $d[\phi(t, x_0), \phi(t, y_0)] < \epsilon$, $\forall t \geq 0$. This is because the point y_0 lies on a cycle different from Γ and, in the nonlinear case, the orbits starting in x_0 and y_0 take different times to go around their respective closed orbits. Then, however close x_0 and y_0 may be (however small $\delta(\epsilon)$), they will move apart ('out of phase') in time and their distance may become greater than the pre-assigned value ϵ. (An example of this situation is provided by the Lotka–Volterra–Goodwin model, see the appendix, p. 128.) This possibility has motivated the following definition regarding an autonomous system of differential equations

$$\dot{x} = f(x) \qquad x \in \mathbb{R}^m. \tag{4.3}$$

definition 4.7 *An orbit generated by system (4.3), with initial condition x_0 on a compact, ϕ-invariant subset A of the state space, is said to be* **orbitally stable** *(***asymptotically orbitally stable***) if the invariant set*

$$\Gamma = \{\phi(t, x_0) | x_0 \in A; t \geq 0\}$$

(the forward orbit of x_0) is stable (asymptotically stable) according to definitions 4.5–4.6.

The second or direct method of Lyapunov can also be used to establish stability or asymptotic stability of invariant sets different from individual points. The strategy is the same as for the proof of stability of fixed points.

We seek a scalar-valued function defined and positive definite on a certain neighbourhood of the set under investigation, which plays the role of generalised distance function. We then try to prove that the function is nonincreasing (or even decreasing) in time. *Converse* results ensure that the existence of such functions is not only a sufficient, but also a necessary condition for stability (or asymptotic stability). Here we limit ourselves to stating the following general result for flows on metric spaces (cf. Hahn, 1963, pp. 76–7).

theorem 4.1 *A necessary and sufficient condition for the (Lyapunov) stability of a compact subset $A \subset M$ of the state space, invariant for the flow ϕ generated by (4.3), is the existence of a scalar-valued, continuously differentiable function $V(x)$ defined on an open neighbourhood $N(\alpha, A)$ of A, with the following properties:*

(i) *for every $\epsilon_1 > 0$ there exists a $\delta_1(\epsilon_1) > 0$ such that $V(x) > \delta_1$ if $x \in N(\alpha, A)$ and $d(x, A) > \epsilon_1$;*

(ii) *for every $\epsilon_2 > 0$, there exists a $\delta_2(\epsilon_2) > 0$ such that $V(x) < \epsilon_2$ if $x \in N(\alpha, A)$ and $d(x, A) < \delta_2$;*

(iii) *if $\phi(t, x)$ is the solution of (4.3), then $V[\phi(t, x)]$ is nonincreasing for $t \geq 0$ as long as $\phi(t, x) \in N(\alpha, A)$.*

A necessary and sufficient condition for asymptotic stability of A is that the function $V(x)$ has properties (i)–(iii) above and, in addition

(iv) $\lim_{t \to +\infty} V[\phi(t, x)] = 0$ *for $\phi(t, x) \in N(\alpha, A)$ and $t > 0$.*

4.3 Attracting sets and attractors

The fact that a set is attracting does not mean that all its points are attracting too. The more restrictive notion of 'attractor' is needed. Unfortunately, there is not yet a general consensus on the proper characterisation of attractors and the definitions vary considerably in the specialised literature. However, most authors agree, broadly, that an attractor should have two basic properties:

(i) attractiveness and
(ii) indecomposability.

We shall deal first with property (ii), which is less controversial and is usually characterised by one or the other of the following two definitions.

definition 4.8 *A flow ϕ is called* **topologically transitive** *on an invariant set A if, for any two open sets $U, V \subset A$ there exists a $t \in \mathbb{R}$ such that $\phi(t, U) \cap V \neq \emptyset$. Analogously, a map G is topologically transitive on A if, for any two open sets $U, V \subset A$, there exists an integer $n \in \mathbb{Z}$ such that $G^n(U) \cap V \neq \emptyset$.*

definition 4.9 *A flow ϕ (or a map G) is* **topologically transitive** *on an invariant set A if the orbit of some point $x \in A$, that is, the set $\{\phi(t, x) | t \in \mathbb{R}\}$ (or, for maps, the set $\{G^n(x) | n \in \mathbb{Z}\}$) is dense in A.*[4]

(For noninvertible maps, only positive values of n and forward orbits are considered.) If the set A is compact, the two definitions are equivalent. In either case, we also say that A is a topologically transitive set (for the flow ϕ or the map G), or that A is **indecomposable** in the sense that ϕ (or G) mixes up points of A in such a way that we cannot split it into two dynamically independent pieces.

There is, instead, substantial disagreement on the best way of characterising attractiveness. Here we have two basic alternatives. In its sharpest form, the first alternative is represented by the following definition.

definition 4.10 *An* **attractor** *is a topologically transitive set, asymptotically stable according to definitions 4.5 and 4.6.*

With nonessential differences in formulation, definition 4.10 agrees with the definitions of attractors used, for example, by Guckenheimer and Holmes (1983), pp. 34–6; Wiggins (1990), p. 45; Perko (1991), p. 193; Arrowsmith and Place (1992), pp. 137–8; Katok and Hasselblatt (1995), p. 128; Robinson (1999), pp. 326–7.

This definition of attractor raises some difficult conceptual problems. For example, there exist well-known cases of dynamical systems that do not possess any set satisfying definition 4.10, even though most of their orbits converge asymptotically to a unique set (cf. Milnor, 1985, p. 178). Some known examples are discussed in chapter 6, section 6.7 where we deal with chaotic attractors. These difficulties led to the following less restrictive definition.

definition 4.11 *A compact invariant subset of the state space $A \subset M$ is called* **attractor** *if*

(a) *its basin of attraction, or stable set $B(A) = \{x \in M | \omega(x) \subset A\}$, has strictly positive Lebesgue measure;*[5]

---

[4] A subset B of A is said to be **dense** in A if, for every $x \in A$ and every $\epsilon > 0$, there exists a point $y \in B$ such that $d(x, y) < \epsilon$.

[5] A detailed discussion of the Lebesgue measure can be found in chapter 9.

(b) there is no strictly smaller closed set $A' \subset A$ so that $B(A')$ coincides with $B(A)$ up to a set of Lebesgue measure zero;

or, alternatively,

(b') A is indecomposable in the sense of definitions 4.8–4.9.

This definition is essentially the same as those found, for example, in Milnor (1985), p. 179; Palis and Takens (1993), p. 138; Alligood *et al.* (1997), p. 235. The definition of strange attractor given by Guckenheimer and Holmes (1983), p. 256 also characterises attractiveness as in definition 4.11. The latter authors call definition 4.10 'naive'.

remark 4.3 A few points deserve attention here:

(1) The most important difference between definition 4.10 and definition 4.11 is that the latter

   (a) does not require that A is Lyapunov stable and
   (b) does not require that A attracts a whole neighbourhood.

(2) A set of positive Lebesgue measure satisfying definition 4.11 need not attract anything outside itself (cf. Milnor, 1985, p. 184). This would not suffice to satisfy definition 4.10. We shall see a very well-known example of this case later on in this chapter.

(3) In Milnor's terminology an attractor in the sense of definition 4.10 could be called 'asymptotically stable attractor' (cf. Milnor, 1985, p. 183). On the other hand, Robinson (1999), p. 327, would call an attractor in the sense of definition 4.11 'the core of the limit set'.

(4) In certain definitions, it is required that the stable set is *open*. Of course, open subsets of $\mathbb{R}^m$ have positive k-dimensional Lebesgue measure (i.e. positive length, area, volume, etc.).

Notice that, in view of (1) and (2), an attractor in the sense of definition 4.11 need not be stable under random perturbations of arbitrarily small amplitude. The difference between the stronger definition 4.10 and the weaker definition 4.11 can be appreciated by two simple, textbook examples, one in discrete time the other in continuous time.[6]

Consider the map on the unit circle S^1 defined by

$$\theta_{n+1} = G(\theta_n) = \theta_n + 10^{-1}\sin^2(\pi\theta_n) \mod 1 \qquad (4.4)$$

where θ denotes an angle and the map rotates the circle anticlockwise, the rotation being a nonlinear function of θ and the exact meaning of mod

---

[6]Cf. Katok and Hasselblatt (1995), p. 128; Robinson (1999), p. 327.

(modulo) is explained in remark 4.4 below. The only fixed point of G on S^1 is $\theta = 0$, and $L^+(G) = L^-(G) = \{0\}$. Thus, there exist *open* neighbourhoods N of $\{0\}$ on S^1 such that $\omega(\theta) = \{0\}$ for $\theta \in N$. Consequently, $\{0\}$ is an attractor in the weak sense of definition 4.11. However, the fixed point is clearly not Lyapunov stable, and therefore, not asymptotically stable in the stronger sense of definition 4.6. Consequently, it does not satisfy the requirements of definition 4.10 and is not an attractor in that stronger sense (cf. figure 4.6(a) and 4.6(b)).

A continuous-time analogue of 4.4 is given by the system (4.1)–(4.2) introduced in section 4.1 (with a graphical representation in figure 4.3). The fixed point $(1,0)$ is an attractor in the weak sense of definition 4.11. However, $(1,0)$ is *not* (Lyapunov) stable and therefore not an attractor in the sense of definition 4.10.

remark 4.4 We shall have many occasions to use circle maps, like map (4.4), and we clarify a few basic points in this remark. The unit circle, denoted S^1, can be represented in two different ways:

(1) the **multiplicative notation**

$$S^1 = \{z \in \mathbb{C} \,\big|\, |z| = 1\} = \{e^{i2\pi\theta}, \theta \in \mathbb{R}\}$$

(2) the **additive notation**

$$S^1 = \mathbb{R}/\mathbb{Z} = \{\theta \in \mathbb{R} \mod 1\}$$

where mod 1 means that two real numbers which differ by an integer are identified.[7]

Figure 4.6 is an illustration of map (4.4) on the circle in the additive (a) and multiplicative notations (b). The labels multiplicative and additive are easily understood if we consider that a fixed rotation by an angle $2\pi\alpha$ can be represented by cz $(c = e^{i2\pi\alpha})$ in the former notation and by $\theta + \alpha \mod 1$ in the latter.

The multiplicative and additive notations are related by the map $h(\theta) = e^{i2\pi\theta}$, which establishes an equivalence relation between the two representations called 'isomorphism'. As we shall see in greater detail in chapter 9, this means, roughly speaking, that the two representations have the same properties from a measure-theoretic point of view.

There exist different distance functions satisfying the properties of a metric for the circle (see remark 1.2 in section 1.4), for example:

---

[7]More generally, we have $x \mod p = x - p \operatorname{Int}\left(\frac{x}{p}\right)$ where $p \geq 1$ is an integer and $\operatorname{Int}(\cdot)$ means 'the integer part of'. The reader should also notice the different notations A/B and $A \setminus B$, for the 'quotient space' and 'difference space', respectively.

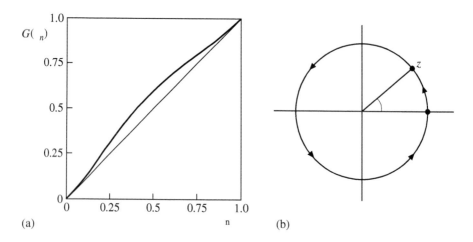

Fig. 4.6 The map (4.4) represented (a) in additive notation; (b) in multiplicative notation.

(1) $d(z_1, z_2) = |z_1 - z_2|$;
(2) $d(z_1, z_2) = 0$ if $z_1 = z_2$;
$\qquad\qquad = \pi$ if z_1 and z_2 are antipodal;
$\qquad\qquad =$ the length of the shorter of the two arcs of S^1 joining z_1
$\qquad\qquad$ and z_2, otherwise.

(Remember that if the length of the radius is one, the arc length is equal to the value of the angle measured in radians.)

The map (4.4) and the maps

$$G(x) = x + \alpha \quad \mathrm{mod}\ 1$$
$$G(x) = 2x \qquad \mathrm{mod}\ 1$$

are *continuous* when we view them as maps of the circle with either of the two metrics above. On the other hand, those maps are *not* continuous as maps of the interval $[0, 1)$ with the Euclidean metric. We shall see, however, that continuity may not be important when we study certain properties of the maps such as Lyapunov characteristic exponents or metric entropy.

4.4 Periodic orbits and their stability

Having thus surveyed some of the basic concepts and definitions necessary to characterise the different types of asymptotic behaviour of nonlinear dynamical systems, we now turn to the analysis of a class of nonconstant solutions of differential and difference equations which is next to fixed points in the scale of complexity, namely periodic orbits. The latter are interesting in themselves as mathematical representation of periodicities in natural and social phenomena. Moreover, as we shall see below, (unstable) periodic orbits are essential ingredients in more complicated orbit configurations, such as chaotic sets.

We begin the discussion with periodic solutions of continuous-time systems.

4.4.1 Periodic orbits of continuous-time systems

Consider, again, the basic system of differential equations (4.3) $\dot{x} = f(x)$ and the derived flow ϕ. A solution $\phi(t, x^*)$ of system (4.3) through a point x^* is said to be **periodic** of (minimum) period $T > 0$ if $\phi(T, x^*) = x^*$. The invariant set $\Gamma = \{\phi(t, x^*) | t \in [0, T)\}$ is a closed curve in the state space and is called a **periodic orbit** or **cycle**. T is called the **period** of the cycle and measures its time-length.

We know that for a system like (4.3), if f is continuously differentiable, there exists a unique solution through any given point. This suggests that the asymptotic behaviour of solutions of differential equations is constrained by the dimension of its state space. Thus, for one-dimensional flows in $\mathbb{R}$, orbits can only converge to a fixed point or diverge from it. Periodic solutions are possible only for two- or higher-dimensional flows and, in fact, periodicity is the most complicated type of asymptotic behaviour of flows on $\mathbb{R}^2$.

Notice also that *isolated* periodic solutions (limit cycles) are possible only for nonlinear differential equations. As we discussed in chapter 2, linear systems of differential equations in $\mathbb{R}^m$ $m \geq 2$, characterised by a pair of purely imaginary eigenvalues (the case of centre) have a *continuum* of periodic solutions, one for every initial condition. Moreover, a limit cycle can be structurally stable in the sense that, if it exists for a given system of differential equations, it will persist under a slight perturbation of the system in the parameter space. On the contrary, periodic solutions of a linear system can be destroyed by arbitrarily small perturbations of the coefficients.

Proving the existence of periodic orbits for systems of ordinary different equations is not always easy. A general local result known as 'Hopf bifurcation' will be studied in chapter 5. For systems in the plane, we can sometimes establish the existence of (one or more) limit cycles by means of the celebrated Poincaré–Bendixson theorem, a version of which follows.

theorem 4.2 (Poincaré–Bendixson) *Consider a system of differential equations $\dot{x} = f(x)$, with $f : U \to \mathbb{R}^2$, U a connected subset of $\mathbb{R}^2$, $f \in C^1$, and let ϕ be the flow generated by the vector field f and x_0 be the initial value. Suppose*

(i) *there exists a closed and bounded region $C \subset U$ such that $\phi(t, x_0) \in C$ for all $t \geq T$ and*

(ii) *there are no fixed points in C.*

Then, there exists at least one periodic solution in C, which is contained in the ω-limit set of x_0.

A proof of the theorem can be found, for example, in Glendinning (1994), pp. 135–6; Katok and Hasselblatt (1995), pp. 452–3; Robinson (1999), pp. 182–3. The basic condition required by the theorem is the existence of a 'trapping region' such that orbits of the system can enter but cannot leave it, and which contains no fixed point. Then the proof relies on the geometrical fact (sometimes referred to as the 'Jordan curve theorem') that in $\mathbb{R}^2$ a closed curve divides the state space into two disjoint parts: 'inside' and 'outside' the curve, so that to move from one part to the other, a path must cross the curve. (This, of course, need not be the case in $\mathbb{R}^m$, $m > 2$.)

Local stability of a periodic solution of system (4.3) can be discussed in terms of eigenvalues of certain matrices. For this purpose, we adopt the representation $\Gamma = \{x^*(t) | t \in [0, T)\}$, $x^*(t) = x^*(t + T)$. Once again we make use of the auxiliary function

$$\xi(t) \equiv x(t) - x^*(t)$$

and linearising $\dot{\xi}(t)$ about $\xi(t) = 0$, i.e., about the periodic orbit Γ, we obtain

$$\dot{\xi} = A(t)\xi \qquad (4.5)$$

where the matrix $A(t) \equiv Df[x^*(t)]$ has periodic coefficients of period T, so that $A(t) = A(t + T)$. Solutions of (4.5) take the general form

$$V(t)e^{\lambda t}$$

where the vector $V(t)$ is periodic in time with period T, $V(t) = V(t + T)$. Let $X(t)$ denote the fundamental matrix, that is, the $m \times m$ time-varying matrix whose m columns are solutions of (4.5). For a T-periodic system like (4.5) $X(t)$ can be written as

$$X(t) = Z(t)e^{tR}$$

where $Z(t)$ is an $(m \times m)$, T-periodic matrix and R is a constant $(m \times m)$ matrix. Moreover, we can always set $X(0) = Z(0) = I$ whence

$$X(T) = e^{TR}.$$

The dynamics of orbits near the cycle Γ are determined by the eigenvalues $(\lambda_1, \ldots, \lambda_m)$ of the matrix e^{TR} which are uniquely determined by (4.5).[8] The λs are called **characteristic (Floquet) roots or multipliers** of (4.5), whereas the eigenvalues of R $(\chi_1, \ldots, \chi_m)$ are called **characteristic (Floquet) exponents**.

One of the roots (multipliers), say λ_1, is always equal to one, so that one of the characteristic exponents, say χ_1, is always equal to zero, which implies that one of the solutions of (4.5) must have the form $V(t) = V(t+T)$. To verify that this is the case put $V(t) = \dot{x}^*(t)$ and differentiating we obtain

$$\frac{d\dot{x}^*(t)}{dt} = Df[x^*(t)]\dot{x}^*(t).$$

Thus $\dot{x}^*(t) = \dot{x}^*(t+T)$ is indeed a solution of (4.5). The presence of a characteristic multiplier equal to one (a characteristic exponent equal to zero) can be interpreted by saying that if, starting from a point on the periodic orbit Γ, the system is perturbed by a small displacement in the direction of the flow, it will remain on Γ. What happens for small, random displacements off Γ depends only on the remaining $(m-1)$ multipliers $\lambda_j, (j = 2, \ldots, m)$ (or the remaining $\chi_j, (j = 2, \ldots, m)$ exponents), providing the modulus of none of them is equal to one (respectively, providing none of them is equal to zero). In particular, we have the following.

(i) If all the characteristic multipliers λ_j $(j = 2, \ldots, m)$ satisfy the conditions $|\lambda_j| < 1$, then the periodic orbit is asymptotically (in fact, exponentially) orbitally stable in the sense of definition 4.7.

(ii) If for at least one of the multipliers, say λ_k, $|\lambda_k| > 1$, then the periodic orbit is unstable.

We can also discuss periodic solutions of system (4.3) and their stability by a different, though related, method, which makes use of a Poincaré or first-return map. The map reduces the analysis of an m-dimensional, continuous-time problem to a $(m-1)$-dimensional, discrete-time problem.

Consider again the periodic orbit Γ of (4.3) and define a δ-neighbourhood of Γ, that is the set of points whose distance from Γ is less than a small

---

[8]The matrix e^{TR} itself is uniquely determined but for a similarity transformation, that is, we can substitute e^{TR} with $P^{-1}e^{TR}P^{-1}$ where P is a nonsingular $(m \times m)$ matrix. This transformation leaves eigenvalues unchanged.

quantity δ. Let us define a **surface of section** P_t which, for fixed t, passes through the point $x^*(t) \in \Gamma$ and is perpendicular to Γ at $x^*(t)$. If we write

$$f[x^*(t)] \equiv g(t)$$

the surface is defined by the property

$$([x(t) - x^*(t)], g(t)) = 0$$

where $(\cdot, \cdot)$ denotes scalar product.

A surface

$$Q_t = N(\delta, \Gamma) \cap P_t$$

(sometimes called a **disk**) is the intersection of the δ-neighbourhood of Γ and the surface of section. If we choose the distance δ small enough, the sets $Q_t(t \in [0, T))$ will be disjoint. At time $t = 0$, let us choose a point $x_0 \in Q_0$ close to the point $x^*(0) \in \Gamma$. Then, if $\|x_0 - x^*(0)\|$ is sufficiently small, under the action of the flow (4.3) the orbit starting at x_0 must return to Q_0 after a time $\tau(x_0) \simeq T$ (and $\tau \to T$ as $\|x_0 - x^*(0)\| \to 0$).

Then the **Poincaré or first-return map** is defined as $G : Q \to P$

$$G(x) = x[\tau(x)] \tag{4.6}$$

$x = \phi_0(x)$ and $x[\tau(x)] = \phi_{\tau(x)}(x)$ where, as usual, $\phi_t(x) = \phi(t, x)$ (see figure 4.7). Equation (4.6) can also be written in the usual way as

$$x_{n+1} = G(x_n)$$

which uniquely defines a sequence of points in the $(m-1)$-dimensional state space $P \subset \mathbb{R}^{m-1}$. However, successive points are separated by intervals of time that are *not* constant but change at each iteration as functions of the state of the system. The properties of G and, in particular, the eigenvalues of the matrix of its derivatives with respect to x do not depend on the specific disk Q_t chosen.

The stability of the periodic orbit can then be ascertained in terms of the eigenvalues of the matrix

$$DG[x^*(t)]$$

the matrix of the derivatives of the Poincaré map[9] evaluated at any point $x^*(t) \in \Gamma$.

A slightly more complicated case occurs when, in a time T (equal to the period of the continuous orbit), the orbit Γ intersects the surface of section

[9] Notice that, in order to construct the matrix $DG(x)$, G must be a differentiable function. This in turn requires that $\tau(x)$ is differentiable. This is generally true if the function f defining the flow is differentiable (cf. Robinson, 1999, theorem 8.1, p. 169).

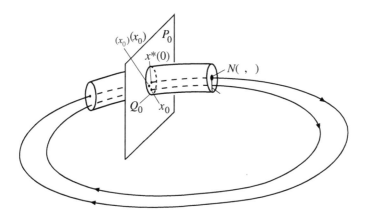

Fig. 4.7 Poincaré map of the periodic orbit of a flow

P_t transversally a number of times $k > 1$ (see figure 4.8). As mentioned before, this occurrence could be avoided by choosing a sufficiently small neighbourhood of Γ (i.e., a sufficiently small δ). However, because the discussion of this occurrence is interesting, we pursue it briefly.

In the multi-intersection case, if we choose an initial point $x_0 \in Q_0$, $x_0 \notin \Gamma$ but sufficiently near a point $x^*(0) \in \Gamma$, the orbit will return to Q_0 k times, after intervals equal to

$$\{\tau_1 [x^*(0)], \tau_2 [\phi_{\tau_1} [x^*(0)]], \ldots, \tau_k [\phi_{\tau_{k-1}} [x^*(0)]]\}$$

where

$$\sum_{i=1}^{k} \tau_i \equiv \tilde{\tau} \simeq T,$$

T being the period of the orbit Γ, and $\tilde{\tau} \to T$ as $\|x_0 - x^*(0)\| \to 0$. We can then define a Poincaré map

$$G^k : Q \to P$$
$$G^k(x) = x[\tilde{\tau}(x)].$$

The stability properties of Γ can be once again established in terms of the eigenvalues of the matrix

$$DG^k(x_i^*) = DG(x_{i-1}^*)DG(x_{i-2}^*) \ldots DG(x_{i-k}^*),$$

evaluated at any of the k intersection points $x_i^* \in \Gamma \cap P$ (i an integer $0 \leq i \leq k - 1$), if none of the eigenvalues is equal to one in modulus. The $(m - 1)$ eigenvalues of $DG^k(x_i^*)$ are the same independently of the choice of

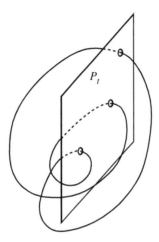

Fig. 4.8 A periodic orbit of period three.

i and are equal to the characteristic multipliers $(\lambda_2, \ldots, \lambda_m)$ of the orbit Γ ($|\lambda_1|$ being equal to one).

Rigorous analytical methods for constructing Poincaré surfaces of section and maps and to calculate the characteristic multipliers are not available generally, that is, for given systems of differential equations. In applications, therefore, one must rely on numerical methods.

remark 4.5 It should be noted that the phrase 'characteristic (Floquet) roots or multipliers' of a periodic orbit is used here to indicate all the m eigenvalues of the matrix e^{TR}. The phrase is sometimes used to refer to the $(m-1)$ eigenvalues of the matrix of derivatives of the Poincaré map.

4.4.2 Periodic orbits of maps

For the usual, general system of difference equations

$$x_{n+1} = G(x_n) \qquad x_n \in \mathbb{R}^m \tag{4.7}$$

a point x^* is periodic of (minimal) period k if $x^* = G^k(x^*)$ and $G^n(x) \neq x$ for $1 \leq n < k$, where G^k denotes the kth iterate of the map G. Thus, the orbit $\{x^* = G^k(x^*), G^{k-1}(x^*), \ldots, G(x^*)\}$ is a sequence of k distinct points that, under the action of G are visited repeatedly by the system, always in the same (reversed) order.

A periodic solution of period k for G corresponds to k distinct fixed points for the map $F \equiv G^k$. In principle, therefore, the properties of periodic solutions of a map can be studied in terms of the corresponding properties of the fixed points of a related (more complicated) map.

Lyapunov stability and asymptotic stability of a periodic point can be defined in a manner analogous to the case of a fixed point.[10] In particular, local asymptotic stability of a periodic point of period k of a map G can be verified by considering the system

$$x_{n+1} = F(x_n) \qquad (4.8)$$

and any one of the k solutions of the equation $\bar{x} = F(\bar{x})$ that correspond to the k periodic points of G. (Notice that $\bar{x} = F(\bar{x})$ is also solved, trivially, by the *fixed* points of G.) Linearising (4.8) around any of the solutions $\bar{x}$ and putting $\xi_n \equiv x_n - \bar{x}$, we have

$$\xi_{n+1} = B\xi_n \qquad (4.9)$$

where, by the chain rule

$$B \equiv DF(\bar{x}) = DG^k(\bar{x}) = DG\left[G^{k-1}(\bar{x})\right]DG\left[G^{k-2}(\bar{x})\right]\cdots DG(\bar{x}) \quad (4.10)$$

where $DG(x)$ denotes the matrix of partial derivatives of G, evaluated at the point x. In (4.10), B is the product of the matrices DG evaluated at the k periodic points of the cycle.

If the matrix B has no eigenvalues with unit modulus (and therefore $\bar{x}$ is hyperbolic), then near $\bar{x}$ the nonlinear map (4.8) is topologically conjugate to the linear approximation (4.9). Then if all eigenvalues of B have modulus less than one, $\bar{x}$ is a locally asymptotically stable fixed point for the map G^k and a locally asymptotically stable periodic point for the map G.

Notice that, for a periodic orbit of period k, $\{x_0^*, x_1^*, \ldots, x_{k-1}^*\}$, where $x_j^* = G^j[x^*(0)]$ and $G^k(x_j^*) = x_j^*$, $(j = 0, 1, \ldots, k-1)$, the matrices of partial derivatives $DG^k(x_j^*)$ have the same eigenvalues for all j. Therefore the stability properties of the k periodic points are the same. Hyperbolicity of a periodic point x_j^* of period k can be defined in a manner perfectly analogous to that of a fixed point. Thus, we say that x_j^* is **hyperbolic** if all the eigenvalues κ of the matrix $DG^k(x_j^*)$ satisfy $|\kappa| \neq 1$.

remark 4.6 For a one-dimensional, discrete-time system

$$x_{n+1} = G(x_n) \qquad x_n \in \mathbb{R}$$

periodic orbits of period $k > 1$ are possible only if:

[10]Notice that for a map G an eventually periodic point x^* is a point that converges to a periodic orbit in finite time. Formally we have: a point x^* is **eventually periodic** of period k for a map G, if there exists an integer $j > 0$ such that $G^{j+k}(x^*) = G^j(x^*)$ and therefore, $G^{i+k}(x^*) = G^i(x^*)$ for $i \geq j$.

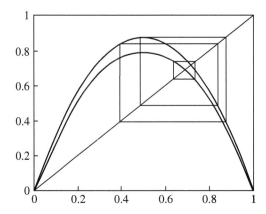

Fig. 4.9 Periodic orbits: lower curve, period-2 cycle; upper curve, period-4 cycle

(1) G is monotonic (and therefore invertible) and $G'(x) < 0$; or
(2) G is noninvertible.

Examples of period-2 and period-4 cycles are given in figure 4.9, for the noninvertible logistic map $x_{n+1} = \mu x_n(1 - x_n)$ for $\mu = 3.2$ and $\mu = 3.5$.

4.5 Quasiperiodic orbits

The next level of complexity is given by systems generating aperiodic orbits whose dynamics is relatively simple and not chaotic in the sense to be discussed in chapter 6. We call those orbits quasiperiodic. The essential character of quasiperiodic dynamics can be easily understood by employing once again, a map of a circle. Making use of the multiplicative notation discussed in remark 4.4, we write

$$G_C : S^1 \to S^1 \qquad z_{n+1} = G_C(z_n) = cz_n \qquad (4.11)$$

with $z = e^{i2\pi\theta}$, $\theta \in \mathbb{R}$, α a positive constant, $c = e^{i2\pi\alpha}$. Thus, (4.11) describes a fixed, anticlockwise rotation of the circle by an angle α. If α is rational, i.e., $\alpha = p/q$, p and q integers, then any initial point on S^1 is periodic of period q for the map G_C. If α is irrational, at each iteration (rotation) a new point is reached on the unit circle, but no point is ever revisited in finite time. The system will get arbitrarily close, again and again, to any point of the circle, that is to say, the map G_C is topologically transitive (the orbit of any $x \in S^1$ is dense in S^1). However, the images of close points remain close and we do not have the phenomenon of the

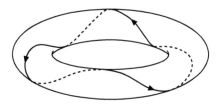

Fig. 4.10 Periodic orbit on a 2-dimensional torus

separation of nearby orbits that, as we shall see, is a typical feature of chaos.

In continuous time, the simplest case of quasiperiodic dynamics is given by a system defined by a pair of oscillators of the form

$$\ddot{x} + w_1^2 x = 0, \qquad \ddot{y} + w_2^2 y = 0 \tag{4.12}$$

where $x, y \in \mathbb{R}$ and w_1, w_2 are real constants. System (4.12) can be rewritten as the following first-order linear system of differential equations in $\mathbb{R}^4$:

$$\begin{aligned}
\dot{x}_1 &= -w_1 x_2 & \dot{y}_1 &= -w_2 y_2 \\
\dot{x}_2 &= w_1 x_1 & \dot{y}_2 &= w_2 y_1.
\end{aligned} \tag{4.13}$$

The unique fixed point of (4.13) is the origin of the coordinate axes and is a centre. Transforming the variables x_1, x_2 and y_1, y_2 into polar coordinates, system (4.13) can be written as

$$\begin{aligned}
\dot{\theta}_1 &= w_1 & \dot{r}_1 &= 0 \\
\dot{\theta}_2 &= w_2 & \dot{r}_2 &= 0
\end{aligned} \tag{4.14}$$

where θ_i denotes the angle and r_i the radius. For a given pair $(\bar{r}_1, \bar{r}_2)$, $r_i > 0$ $(i = 1, 2)$, the state space thus redefined is known as a two-dimensional torus (see figure 4.10).

For the dynamics of system (4.14) on the torus, there exist two basic possibilities:

(i) w_1/w_2 is a rational number, i.e., it can be expressed as a ratio of two integers p/q, in which case there is a continuum of periodic orbits of period q;

(ii) w_1/w_2 is an irrational number, in which case starting from any initial point on the torus, the orbit wanders on it, getting arbitrarily near any other point, without ever returning to that exact initial point. The flow generated by (4.14) is topologically transitive on the torus;

(iii) in both cases (i) and (ii), close points remain close under the action of
the flow.

If we want to describe an *attracting* quasiperiodic set we need to modify
system (4.14), for example,

$$\dot{\theta}_1 = \omega_1 \qquad \dot{r}_1 = r_1(1 - r_1^2)$$
$$\dot{\theta}_2 = \omega_2 \qquad \dot{r}_2 = r_2(1 - r_2^2). \tag{4.15}$$

In this case, for the flow (4.15) there exists an invariant set $T^2 = S^1 \times S^1$, a
2-torus which attracts all the orbits (except those starting at $r_1 = r_2 = 0$,
the trivial, unstable fixed point). If ω_1/ω_2 is irrational, the dynamics on T^2
is quasiperiodic.

A general definition of quasiperiodicity of an orbit as a function of time
can be given as follows:

definition 4.12 *A function $h : \mathbb{R} \to \mathbb{R}^m$ is called* **quasiperiodic** *if it can
be written in the form $h(t) = H(\omega_1 t, \omega_2 t, \ldots, \omega_m t)$, where H is periodic of
period 2π in each of its arguments, and two or more of the m (positive)
frequencies are incommensurable.*

Thus, we call quasiperiodic a continuous-time system whose solution
$\phi(t, x)$ for a typical initial point x_0, is a quasiperiodic function of time.
Quasiperiodic orbits can look quite complicated, especially if there exist
many incommensurable frequencies. As a matter of fact, quasiperiodic dynamics on hyper-tori, e.g., T^n, $n \geq 3$, are rarely observed because even slight
changes to their dynamics perturb the quasiperiodic set into a chaotic set.
We shall return to this point in chapter 8.

Appendix: conservative and dissipative systems

A characterisation of dynamical systems related to the notions of stability
and attractiveness is that of conservative and dissipative systems. A system
is said to be **conservative** if volumes in the state space are kept constant
as time evolves. In continuous-time, let us consider the usual system of
ordinary differential equations

$$\dot{x} = f(x) \qquad x \in \mathbb{R}^m \tag{4.16}$$

and let us choose at time $t = 0$ a closed $(m-1)$-dimensional surface $S(t)$ in
$\mathbb{R}^m$ with a m-dimensional volume $V(0)$. Then the evolution in time of V is

given by

$$\frac{dV(t)}{dt} = \int \sum_{i=1}^{m} \frac{\partial f_i(x)}{\partial x_i} dx \qquad (4.17)$$

where the integral is taken over the volume $V(t)$ interior to the surface $S(t)$. We say that system (4.16) is **conservative** if $dV(t)/dt = 0$, that is, if the volume $V(t)$ remains constant in time. Notice that the expression $\sum_{i=1}^{m} \partial f_i(x)/\partial x_i$ denotes the trace of the time-varying matrix $Df[x(t)]$.

Analogously, a discrete-time system $x_{n+1} = G(x_n)$, $x_n \in \mathbb{R}^m$ is said to be conservative if $|\det DG(x_n)| = 1$. From the fact that in conservative systems volumes remain constant under the flow (or map), we may deduce that those systems cannot have asymptotically stable sets. Nevertheless, conservative systems may be characterised by complex or (suitably defined) chaotic behaviour.

In this book we do not study conservative systems, but we would like to mention here one interesting (and rare) economic application. The model was developed by Goodwin (1967), who applied certain equations, previously studied by the mathematical biologist Lotka and the mathematician Volterra, to a classical problem of cyclical economic growth. The formal model is:[11]

$$\dot{x} = \alpha x - \beta xy$$
$$\dot{y} = -\gamma y + \delta xy$$

where $x, y \in \mathbb{R}$ and $\alpha, \beta, \gamma, \delta$ are real, positive constants. In Goodwin's interpretation, x denotes the level of employment (relative to population) and y denotes the share of wages in the national income. Thus, the model can be used to investigate the dynamics of production and the distribution of income. By simple manipulations, one can verify that, along the orbits of the system, the area defined by

$$F(x, y) = e^{-\delta x} e^{-\beta y} x^\gamma y^\alpha$$

is constant and conclude that the system is conservative. It can also be shown the the system has two nonnegative fixed points, the origin (a saddle point) and a second one located in the positive quadrant which is a centre. All the orbits starting in the positive quadrant are closed orbits around the fixed point. There are no attracting limit cycles, and initial conditions entirely determine which of the infinite number of cycles is actually followed.

---

[11]The mathematical system is the same as that encountered in exercises 3.4(g) and 3.7(a) at the end of chapter 3.

Any two-dimensional set of initial conditions is transformed by the action of the flow into another set of equal area.

Unlike conservative systems, **dissipative** dynamical systems are characterised by contraction of state space volumes. Because of dissipation, the dynamics of a system whose state space is m-dimensional, will eventually be confined to a subset of fractal dimension[12] smaller than m (and possibly zero). Thus, in sharp contrast to the situation encountered in conservative systems, dissipation permits one to distinguish between *transient* and *permanent* behaviour. For dissipative systems, the latter may be quite simple even when the number of state space variables is very large.

To better understand this point, think, for example, of an m-dimensional system of differential equations characterised by a unique, globally asymptotically stable equilibrium point. Clearly, for such a system, the flow will contract any m-dimensional set of initial conditions to a zero-dimensional final state, the equilibrium point. Think also of an m dimensional $(m \geq 2)$ dissipative system characterised by a unique, globally stable limit cycle. Here, too, once the transients have disappeared, we are left with a cycle of dimension one.

Exercises

4.1   With reference to figure 4.4, determine the ω-limit set (a point) of the points C, D, E.

4.2   With reference to exercise 3.2(a)–(e) at the end of chapter 3, for each of the one-dimensional flows determine the ω- and α-limit sets of points on the real line.

4.3   Consider the system (4.1)–(4.2)

$$\dot{r} = r(1 - r^2)$$
$$\dot{\theta} = \theta^2(2\pi - \theta)^2.$$

Discuss the α- and ω-limit sets of points in different regions of $\mathbb{R}^2$.

4.4   For the following systems (where again r denotes the radius and θ the angle) find the nonwandering set $\Omega(\phi)$ and the positive limit set $L^+(\phi)$

(a)

$$\dot{r} = r(1 - r^2)$$
$$\dot{\theta} = 1$$

---

[12]The concept of fractal dimension will be discussed in chapter 7.

(b)

$$\dot{r} = r(1 - r^2)$$
$$\dot{\theta} = (\pi - \theta)^2.$$

4.5   Determine $w(r_0, \theta_0)$ of the systems below for the given initial values

(a) (r_0, θ_0): $(0,0)$, $(1/2, 0)$, $(1,0)$, $(3/2, 0)$, $(2,0)$, $(3,0)$

$$\dot{r} = r(r - 1)(2 - r)$$
$$\dot{\theta} = 1$$

(b) (r_0, θ_0): (r_0^1, θ_0), (r_0^2, θ_0), (r_0^3, θ_0) where $r_0^1 < 1$, $r_0^2 = 1$, $r_0^3 > 1$ and $0 \le \theta < 2\pi$

$$\dot{r} = r(r^2 - 1)$$
$$\dot{\theta} = (\pi - \theta)^2.$$

4.6   Prove that each of the following subsets of $\mathbb{R}^2$ is invariant for the corresponding flow:

(a) The open set $\{(x, y) \in \mathbb{R}^2 | y > 0\}$ for the system

$$\dot{x} = 2xy$$
$$\dot{y} = y^2.$$

(b) The closed set formed by: (i) the curve beginning at $(x, y) = (1, 0)$ and completing a cycle of 2π radians of the spiral defined in polar coordinates as $r = e^{-\theta}$ (r the radius, θ the angle in radians) and (ii) the segment obtained by joining the point $(x, y) = (e^{-2\pi}, 0)$ and the point $(x, y) = (1, 0)$, for the system

$$\dot{x} = -x - y$$
$$\dot{y} = x - y.$$

(*Hint*: transform the system in polar coordinates, then verify the possibility that orbits with initial conditions belonging to the closed set ever leave it.)

4.7   Use the Poincaré–Bendixson theorem to show that, for the given system, there exists an invariant set in $\mathbb{R}^2$ containing at least one limit cycle:

(a) (Robinson, 1999, p. 182, example 9.1)

$$\dot{x} = y$$
$$\dot{y} = -x + y(1 - x^2 - 2y^2).$$

(b) (Glendinning, 1994, p. 136, example 5.10)

$$\dot{x} = y + \frac{1}{4}x(1 - 2r^2)$$

$$\dot{y} = -x + \frac{1}{2}y(1 - r^2)$$

which has a single fixed point at the origin. (*Hint*: use polar coordinates and recall that $\sin^2 \theta = \frac{1}{2}(1 - \cos 2\theta)$, $\cos^2 \theta = \frac{1}{2}(1 + \cos 2\theta)$.)

(c)

$$\dot{z} = Az - r^2 z$$

where $z = (x, y) \in \mathbb{R}^2$, A is a (2×2) positive definite constant matrix and $r^2 = (x^2 + y^2)^{\frac{1}{2}}$. Use Lyapunov's direct method to find the invariant set.

4.8   Prove that the stability properties of each of the k periodic points of a map G are the same.

4.9   Consider the following maps. Determine the fixed points and the periodic points of period 2 and discuss their stability:

(a) $G(x_n) = \mu - x_n^2$, $\mu = 1$
(b) $G(x_n) = \mu x_n(1 - x_n)$, $\mu = 3.2$ (as illustrated in figure 4.9).

4.10   Demonstrate the derivation of (4.14) from (4.12).

4.11   Select values for the parameters w_1, w_2 in system (4.13) that lead to (i) a continuum of periodic orbits on the torus, and (ii) quasiperiodic orbits on the torus.

4.12   Consider the Lorenz model

$$\dot{x} = -\sigma x + \sigma y$$

$$\dot{y} = -xz + rx - y$$

$$\dot{z} = xy - bz$$

where $x, y, z \in \mathbb{R}$; $\sigma, r, b > 0$. Show that the system is dissipative. Find a trapping region including all the fixed points. (*Hint*: define a new variable $u = z - (r + \sigma)$, transform the system in an equivalent system in the variables (x, y, u) and find a Lyapunov function defining a closed region in $\mathbb{R}^3$ such that orbits may enter but never leave it.)

5
Local bifurcations

Bifurcation theory is a very complex research programme which was originally outlined by Poincaré and, in spite of very significant progress in the recent past, is still far from complete. Indeed, even terminology is not uniform in the literature. In this book, we shall introduce only the basic concepts and shall discuss some simple cases, with a view to applications, leaving aside many deeper theoretical questions.

5.1 Introduction

Previous chapters have concentrated almost exclusively on the dynamical properties of *individual* systems of differential or difference equations. In this chapter, we shall instead consider *families* of systems whose members are identical except for the value of one parameter, that is, equations such as

$$\dot{x} = f(x; \mu) \qquad \mu \in \mathbb{R} \tag{5.1}$$

or

$$x_{n+1} = G(x_n; \mu) \qquad \mu \in \mathbb{R}. \tag{5.2}$$

Suppose now that a certain property (e.g., the number of equilibria or their stability) of system (5.1) (or (5.2)), holds for μ in an open interval (a, b), but this property does not hold on any other larger interval. The endpoints a and b are called **bifurcation points**.

Bifurcations can be classified as local, global and local/global. We have a **local** bifurcation when the qualitative changes in the orbit structure can be analysed in a neighbourhood of a fixed or a periodic point of a map, or a fixed point of a flow. (Bifurcations of periodic points of flows can be similarly treated by means of Poincaré maps.) **Global** bifurcations are characterised instead by changes in the orbit structure of the system which are not accompanied by changes in the properties of its fixed or periodic points. **Local/global** bifurcations occur when a local bifurcation has also global repercussions that qualitatively change the orbit structure far from the fixed point. In what follows, we shall almost exclusively consider local bifurcations of flows and maps. The subject of global bifurcations will be taken up again in chapter 8, when we discuss transition to chaos.

Local bifurcations of fixed points are closely related to the (loss of) hyperbolicity of these points. When a system is subject to small perturbations in the parameter space (i.e., a parameter value is slightly changed), near each hyperbolic fixed point of the unperturbed system there will be a hyperbolic point of the perturbed system with the same properties. In particular, the Jacobian matrix of the perturbed system, evaluated at the 'new' fixed point will have the same eigenvalue structure as the Jacobian matrix of the 'old' fixed point. That is to say: for flows, it will have the same number of eigenvalues with positive and negative real parts; for maps, the same number of eigenvalues inside and outside the unit circle in the complex plane. Thus, when we look for local bifurcations of fixed points, we shall try to find those parameter values for which hyperbolicity is lost, i.e., for which the real part of an eigenvalue goes through zero (flows) or its modulus goes through one (maps).

5.2 Centre manifold theory

Local bifurcations depending on one parameter essentially involve one real eigenvalue or a pair of complex conjugate eigenvalues. It is then possible to characterise them completely in a reduced one- or two-dimensional space, even when the phenomenon takes place in a multi-dimensional state space. The most general approach for performing this reduction is based on the notion of a *centre manifold*. Centre manifold theory involves many difficult technical details. For our present purposes, a concise description of the method will suffice. Our presentation follows Guckenheimer and Holmes (1983), section 3.2 and Glendinning (1994), chapter 8, section 8.1, where the reader can find more detailed discussions and examples.

Keeping in mind our review of linear systems of differential equations in chapter 2, let us consider the system

$$\dot{x} = Ax \qquad x \in \mathbb{R}^m.$$

The eigenvalues λ of A can be divided into three sets σ_s, σ_u, σ_c, where $\lambda \in \sigma_s$ if $\mathrm{Re}\,\lambda < 0$, $\lambda \in \sigma_u$ if $\mathrm{Re}\,\lambda > 0$ and $\lambda \in \sigma_c$ if $\mathrm{Re}\,\lambda = 0$. As we know the state space can be split into three invariant subspaces called, respectively, stable, unstable and centre eigenspaces, denoted by E^s, E^u and E^c.

Stable, unstable and centre eigenspaces can be defined analogously for systems of linear difference equations (maps) such as

$$x_{n+1} = Bx_n \qquad x_n \in \mathbb{R}^m$$

with the obvious difference that the allocation of eigenvalues depends on whether their moduli are greater, less than or equal to one.

In chapter 3 we mentioned that the stable and unstable manifolds of a nonlinear system in a neighbourhood of a hyperbolic fixed point are related to the corresponding eigenspaces of the linearised system. The following theorem states that result in a more precise manner, and extends it to the centre manifold.

theorem 5.1 (centre manifold of fixed points) *Consider the system of differential equations*

$$\dot{w} = f(w) \qquad w \in \mathbb{R}^m \tag{5.3}$$

where $f \in C^r$ and $f(0) = 0$. Let E^s, E^u and E^c denote, respectively, the stable, unstable and centre eigenspaces of the matrix $A \equiv Df(0)$ (the Jacobian matrix evaluated at $w = 0$). Then there exist C^r stable and unstable invariant manifolds W^s and W^u tangent to E^s and E^u at $w = 0$, and a C^{r-1} centre invariant manifold tangent to E^c at $w = 0$. W^s and W^u are unique, but W^c is not necessarily so. (If $f \in C^\infty$, then a C^r centre manifold exists for any finite r.)

From theorem 5.1 (henceforth referred to as centre manifold theorem) it follows that, at a bifurcation point where the fixed point is not hyperbolic and W^c is not empty, in a neighbourhood of $w = 0$ system (5.3) is topologically equivalent to

$$
\begin{aligned}
\dot{x} &= H(x) & x \in W^c \\
\dot{y} &= -By & y \in W^s \\
\dot{z} &= Cz & z \in W^u
\end{aligned}
$$

where x, y and z are appropriately chosen local coordinates and B, C are positive definite matrices. Locally, the time evolution of the variables y and z is governed by two *linear* subsystems of differential equations with equilibria in $y = 0$ and $z = 0$, respectively. Recalling that the eigenvalues of a positive definite matrix have positive real parts, we can conclude that, for initial points on W^s or W^u, the motion will converge asymptotically to equilibrium for $t \to +$ or $-\infty$, respectively. In order to complete the study of the dynamics near equilibrium, we need only investigate the reduced nonlinear subsystem

$$\dot{x} = H(x) \qquad x \in W^c \tag{5.4}$$

whose dimension is equal to the number of eigenvalues of the matrix $A \equiv Df(0)$ with zero real part. In most cases we cannot hope to calculate (5.4) exactly but, under certain regularity assumptions on W^c, its dynamics can be approximated arbitrarily closely by means of power series expansions around $\bar{x} = 0$. Also the description of the qualitative properties of each type of local bifurcation can be simplified by certain coordinate transformations that reduce the analytical expression of the vector field on the centre manifold to the simplest form (sometimes called normal form equation) exhibiting those properties.

Suppose now that the vector field f of theorem 5.1 depends on a parameter vector $\mu \in \mathbb{R}^k$ and that a bifurcation occurs for $\mu = 0$. To investigate the behaviour of families of systems depending on μ near the bifurcation point we need to extend the basic vector field and its centre manifold so as to include a *variable* μ. This is done by adding the trivial equation $\dot{\mu} = 0$ with obvious solution $\mu = $ constant. The extended centre manifold $\tilde{W}^c$ has dimension equal to $\dim(E^c) + k$ and is tangent to $E^c(0)$ $(E^c(0) = \{(x,y,z)|y = 0, z = 0\})$ at $x = y = z = 0; \mu = 0$. The *extended* centre manifold and the dynamics on it can be approximated in a neighbourhood of $(x = y = z = 0; \mu = 0)$ by the same procedure as for the *nonextended* vector field. The resulting reduced system has the form

$$F : \tilde{W}^c \to \tilde{W}^c$$
$$\begin{pmatrix} \dot{x} \\ \dot{\mu} \end{pmatrix} = F(x, \mu) = \begin{pmatrix} G(x, \mu) \\ 0 \end{pmatrix}.$$

A bifurcation depending on parameters is called a **codimension-k bifurcation**, where k is the smallest number for which that bifurcation occurs in a persistent way for a k-parameter family of systems.

To understand this idea consider, for example, a continuous-time dynamical system in $\mathbb{R}^m$ like (5.1), with a parameter space (μ_1, μ_2, μ_3), as in figure 5.1(a). In general, the locus of points in the parameter space in which a certain condition is satisfied (e.g., the condition that one eigenvalue, or the real part of a pair of complex conjugate eigenvalues of the Jacobian matrix becomes zero), is a two-dimensional surface $F_1(\mu_1, \mu_2, \mu_3) = 0$. On the other hand, a one-parameter family of systems is represented by a one-dimensional curve γ in the parameter space. A transversal intersection of the curve γ

and the surface $F_1 = 0$ is persistent in the sense that it cannot be perturbed away by slightly changing γ. Accordingly, a bifurcation corresponding to $F_1 = 0$ is generic for a one-parameter family of systems and we call it a codimension-1 bifurcation.

Suppose now a second condition is required (e.g. that another eigenvalue, or the real part of another pair of complex conjugate eigenvalues of the Jacobian matrix becomes zero). This condition will define a second, two-dimensional surface $F_2(\mu_1, \mu_2, \mu_3) = 0$ in the parameter space and the simultaneous verification of $F_1 = 0$ and $F_2 = 0$ typically defines a one-dimensional curve ζ in $\mathbb{R}^3$. The transversal intersection of two curves γ and ζ in $\mathbb{R}^3$ is not generic and can be perturbed away by the slightest perturbation. Thus, a bifurcation requiring both conditions $F_1 = 0$ and $F_2 = 0$ can only occur in a persistent way in two-parameter families of systems characterised by two-dimensional surfaces in the parameter space. We call those bifurcations codimension-2 bifurcations. Higher-order codimension bifurcations can be similarly described. (It should be observed that what is relevant here is the number of controlling parameters, not the dimension of the state space.)

In what follows, we shall discuss the most common types of local codimension-1 bifurcations. This will be accomplished by considering the simplest functional forms satisfying certain restrictions associated with each type of bifurcation.

remark 5.1 Notice that the local stable and unstable manifolds of a fixed point $\bar{x}$ of a vector field $\dot{x} = f(x)$ can be characterised by the stability properties of orbits, as follows:

$$W^s(\bar{x}) = \{x \in N \mid d[\phi(t, x), \bar{x}] \to 0 \text{ as } t \to +\infty \text{ and } \phi(t, x) \in N \; \forall t \geq 0\}$$
$$W^u(\bar{x}) = \{x \in N \mid d[\phi(t, x), \bar{x}] \to 0 \text{ as } t \to -\infty \text{ and } \phi(t, x) \in N \; \forall t \leq 0\}$$

where N denotes a neighbourhood of $\bar{x}$ and, as usual, $\phi(t, x)$ is the flow generated by the vector field, d is a distance function; $t \in \mathbb{R}$. On the contrary, convergence to the fixed point (or divergence from it) of the orbits on the centre manifold W^c cannot be specified without further investigation.

Similar considerations can be developed for discrete-time dynamical systems characterised by **diffeomorphisms** (smooth, invertible maps). In particular, there exists a centre manifold theorem for diffeomorphisms entirely analogous to theorem 5.1 except that the eigenvalues of the Jacobian matrix at equilibrium are split according to whether their moduli are greater, less than or equal to one. We can also define local stable and unstable manifolds of a fixed point $\bar{x} = G(\bar{x})$ of a diffeomorphism G, in terms of their stability

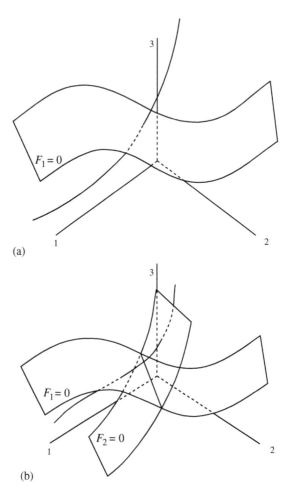

(a)

(b)

Fig. 5.1 Bifurcations: (a) codimension-1; (b) codimension-2

properties, as follows:

$$W^s(\bar{x}) = \{x \in N \mid \lim_{n \to \infty} d[G^n(x), \bar{x}] = 0 \text{ and } G^n(x) \in N \; \forall n \geq 0\} \quad (5.5)$$
$$W^u(\bar{x}) = \{x \in N \mid \lim_{n \to -\infty} d[G^n(x), \bar{x}] = 0 \text{ and } G^n(x) \in N \; \forall n \leq 0\} \quad (5.6)$$

where, again, N is a neighbourhood of $\bar{x}$, $n \in \mathbb{Z}$.

The situation is more complicated, of course, if G is noninvertible, as is the case for many examples discussed in this book. The local **stable** manifold of $\bar{x}$ can be defined exactly as in (5.5), but if G is not invertible, the function G^{-1} is not defined and the unstable manifold of $\bar{x}$ cannot be obtained trivially from (5.5) by simply replacing forward with backward iterates of G. We need the following, more complicated definition to replace (5.6)

$$W^u(\bar{x}) = \{x \in N \mid \text{ there exists a sequence } \{x_i\}_0^\infty$$
$$\text{s.t. } x_0 = x, \ G(x_i) = x_{i-1} \ \forall i \geq 1 \ \text{ and } \ \lim_{i \to \infty} d(x_i, \bar{x}) = 0\}. \tag{5.7}$$

That is, if G is not invertible, each point x_0 in the state space has many 'past histories' and in order to define the unstable manifold of a fixed point, we need to choose those of them that, if 'read backward', converge asymptotically to $\bar{x}$.

5.3 Local bifurcations for flows

After these rather general considerations, in the following pages we discuss the most typical codimension-one bifurcations for flows. For this purpose, the one-parameter family of functions $f(x; \mu)$ will be treated as a function of two variables, x and μ.

5.3.1 Saddle-node or fold bifurcation

We begin with the simplest differential equation exhibiting this type of bifurcation, the one-parameter family of differential equations

$$\dot{x} = f(x; \mu) = \mu - x^2 \qquad x, \mu \in \mathbb{R}. \tag{5.8}$$

The equilibrium solutions of (5.8), as functions of μ, are

$$\bar{x}_{1,2} = \pm\sqrt{\mu}. \tag{5.9}$$

Thus, system (5.8) has two real fixed points for $\mu > 0$. As μ decreases through zero and becomes negative, the two fixed points coalesce and disappear. Thus, a qualitative change in the dynamical properties of (5.8) occurs at $\mu = 0$, which is called a **saddle-node** or **fold bifurcation**. (In a one-dimensional system such as (5.8), it is also known as a **tangent bifurcation**.) The situation is depicted in figure 5.2, where we plot (5.8) in the $(\dot{x}, x)$ plane for three values of the parameter, $\mu < 0$, $\mu = 0$ and $\mu > 0$. This information is presented in a different form in figure 5.3, which depicts the equilibrium values as determined by (5.9) in the (μ, x) plane and is called the **bifurcation diagram**. In bifurcation diagrams, such as figure 5.3, stable fixed points are represented as a solid curve (the **stable branch**), whereas unstable fixed points are represented as a dashed curve (the **unstable branch**). Arrows may be included to emphasise convergence

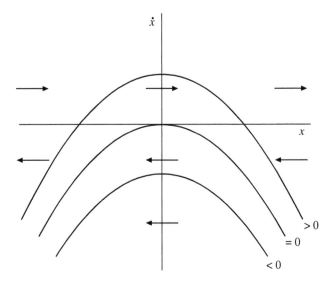

Fig. 5.2 Fold bifurcation for flows

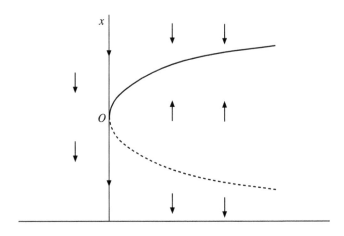

Fig. 5.3 Bifurcation diagram, fixed points against μ

or divergence. (Generally, in computer-generated bifurcation diagrams only the stable branch appears.) In the present case, if $\mu > 0$, $\bar{x}_1$ is the stable equilibrium to which orbits are eventually attracted. For $\mu < 0$ there are no real solutions for $\bar{x}$ and, consequently, no equilibria.

remark 5.2 The name 'saddle-node' derives from the fact that, in the general, multi-dimensional case, the stable branch is composed of nodes, and the unstable branch of saddles.

For a generic function $f(x; \mu)$, $x, \mu \in \mathbb{R}$, the necessary condition for the existence of any local bifurcation is that a fixed point $\bar{x}$ be nonhyperbolic at a critical value of $\mu = \mu_c$. For flows this condition is:

(i)
$$\frac{\partial f(\bar{x}; \mu_c)}{\partial x} = 0.$$

Notice that $\partial f(\bar{x}; \mu_c)/\partial x$ can be thought of as the (1×1) Jacobian matrix from the linear approximation of the system $\dot{x} = f(x; \mu)$, calculated at $\bar{x}$, μ_c. Condition (i) means that the single eigenvalue of the Jacobian matrix calculated at equilibrium is zero at the bifurcation point $\mu = \mu_c$. Two further conditions are required for a fold bifurcation

(ii)
$$\frac{\partial^2 f(\bar{x}; \mu_c)}{\partial x^2} \neq 0$$

(iii)
$$\frac{\partial f(\bar{x}; \mu_c)}{\partial \mu} \neq 0.$$

Condition (ii) ensures that the point is either a maximum or a minimum, but not simply an inflection point. This in turn implies that the curve of fixed points lies entirely to one side of $\mu = \mu_c$. Condition (iii) ensures that, by the implicit function theorem, there exists a unique function $\mu = \mu(x)$ defined sufficiently near μ_c and such that $f[x; \mu(x)] = 0$. It is easily seen that the equation $\dot{x} = \mu - x^2$ satisfies all these conditions: $\partial f(0,0)/\partial x = 0$ and therefore, the eigenvalue at equilibrium is zero; $\partial^2 f(0,0)/\partial x^2 = -2$ and therefore, $\dot{x}$ is at a maximum at the nonhyperbolic fixed point; $\partial f(0,0)/\partial \mu = +1$. Conditions (i)–(iii) are generic, that is, if they are verified for a certain member of the family of system (5.8) (a certain value of μ), arbitrarily near it there will be another member (another, slightly different value of μ) for which those conditions hold too.

The fold bifurcation is particularly important in applications because, on the one hand, it helps define the parametric range for which equilibria exist. On the other hand, if it can be shown that a fold bifurcation occurs at the endpoint of an open interval of values of the controlling parameter, then at least one stable equilibrium exists over that interval.

remark 5.3 The fold is said to be a **discontinuous** or **catastrophic** bifurcation because the bifurcation point corresponds to an endpoint of the stable branch. On the contrary, a bifurcation is **continuous** if there is a continuous path on the stable branch through a bifurcation point in the bifurcation diagram. As we shall see later, this distinction is very important in practice because the changes in the orbit structure following a discontinuous bifurcation can be much more drastic than those occurring in a continuous one (hence the term 'catastrophic' to designate the former). Discontinuous bifurcations are also related to the interesting phenomenon of hysteresis which will be discussed in chapter 8, section 8.3.

5.3.2 Transcritical bifurcation

Consider the system

$$\dot{x} = f(x; \mu) = \mu x - x^2. \tag{5.10}$$

There are two fixed points, $\bar{x}_1 = 0$ (which is independent of μ) and $\bar{x}_2 = \mu$. Because $\partial f(\bar{x}; \mu)/\partial x = \mu - 2\bar{x}$, for $\mu > 0$, $\bar{x}_1$ is locally unstable and $\bar{x}_2$ is locally stable, and vice versa for $\mu < 0$. If $\mu \neq 0$, the two fixed points are hyperbolic, but at $\mu = 0$ they coalesce and hyperbolicity is lost. However, as μ goes through zero, no fixed points appear or disappear, only their stability properties change. This is called a **transcritical bifurcation** and it is *continuous* in the sense explained in remark 5.3.

For a general function $f(x; \mu)$ there are three conditions for a transcritical bifurcation of a fixed point $\bar{x}$ at a critical value of $\mu = \mu_c$. The first two are the same as conditions (i) and (ii) for the fold bifurcation. In order to have *two* curves of fixed points passing through the bifurcation point μ_c, condition (iii) must be replaced by the condition

(iii') $\qquad \dfrac{\partial f(\bar{x}; \mu_c)}{\partial \mu} = 0 \quad \text{and} \quad \dfrac{\partial^2 f(\bar{x}; \mu_c)}{\partial x \partial \mu} \neq 0.$

It is easy to verify that conditions (i), (ii), and (iii') hold for (5.10). The situation is depicted in figure 5.4. In figure 5.4(a) the function f is plotted in the $(x, \dot{x})$ plane for three values $\mu < 0$, $\mu = 0$ and $\mu > 0$. In figure 5.4(b) the bifurcation diagram is plotted in the (μ, x) plane.

5.3.3 Pitchfork bifurcation

The third type of local bifurcation occurs in the one-parameter family of one-dimensional differential equations

$$\dot{x} = f(x; \mu) = \mu x - x^3. \tag{5.11}$$

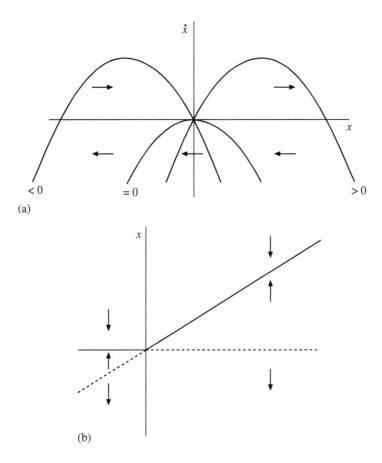

Fig. 5.4 Transcritical bifurcation: (a) equation (5.10); (b) bifurcation diagram

Equilibrium solution $\bar{x}_1 = 0$ is independent of μ. For $\mu > 0$ there are two additional equilibria $\bar{x}_{2,3} = \pm\sqrt{\mu}$. These three equilibria coalesce at $\mu = 0$. Considering that $\dot{x} \gtreqless 0$ if $x(\mu - x^2) \gtreqless 0$ and $\partial f/\partial x = \mu - 3x^2$, we can conclude that, for $\mu < 0$ the unique equilibrium $\bar{x}_1 = 0$ is stable and that, for $\mu > 0$, $\bar{x}_1$ is unstable, and the two nontrivial equilibria are both stable. Therefore at $\mu = 0$ we have a qualitative change in the orbit structure of (5.11). This situation, in which both appearance/disappearance of equilibria and stability changes occur, is called **pitchfork bifurcation** after the shape of the bifurcation diagram. Summarising, a pitchfork bifurcation is characterised by the fact that $x = 0$ is a curve of fixed points and that there exists a second curve passing through $(\bar{x}; \mu_c)$ situated entirely on one side of μ_c. For a generic function $f(x; \mu)$ with a fixed point $\bar{x}$, at a parameter value $\mu = \mu_c$, (i) and (iii') are necessary conditions for a pitchfork bifurcation.

However, (ii) must be replaced by

(ii') $$\frac{\partial^2 f(\bar{x}; \mu_c)}{\partial x^2} = 0 \quad \text{and} \quad \frac{\partial^3 f(\bar{x}; \mu_c)}{\partial x^3} \neq 0.$$

Conditions (ii')–(iii') imply that at the critical point μ_c on the curve of fixed points $\mu(x)$

$$\frac{d\mu(\bar{x})}{dx} = 0 \quad \text{and} \quad \frac{d^2\mu(\bar{x})}{dx^2} \neq 0$$

which guarantees that in the (x, μ) plane, the second curve of fixed points lies entirely on the right side of $\mu = \mu_c$ if the second derivative is positive, on the left side if it is negative.

Conditions (i), (ii') and (iii') are verified for (5.11). At $\mu = 0$ and $\bar{x}(0) = 0$ we have $\partial f/\partial x = 0$, $\partial^2 f/\partial x^2 = 0$, $\partial^2 f/\partial x \partial \mu = 1$, $\partial^3 f/\partial x^3 = -6$. This case is called a **supercritical** pitchfork bifurcation because the nontrivial equilibria appear for $\mu > \mu_c$. This supercritical pitchfork is a continuous bifurcation.

A **subcritical** pitchfork bifurcation occurs if we consider, instead of (5.11), the equation

$$\dot{x} = f(x; \mu) = \mu x + x^3. \tag{5.12}$$

Again, $\bar{x}_1 = 0$ is an equilibrium independently of μ. Two additional nonzero equilibria $\pm\sqrt{-\mu}$ exist for $\mu < 0$. At $\mu = 0$ the three equilibria coalesce at $\bar{x}_1 = \bar{x}_2 = \bar{x}_3 = 0$. The trivial equilibrium is stable for $\mu < 0$ and unstable for $\mu > 0$. The nontrivial equilibria are unstable for (5.12). Conditions (i), (ii'), (iii') are verified in this case, with $\partial^3 f/\partial x^3 = 6$. The term 'subcritical' refers to the fact that the nontrivial equilibria appear for values of the controlling parameter smaller than the bifurcation value. This subcritical pitchfork is a discontinuous bifurcation. A supercritical pitchfork bifurcation is illustrated in figure 5.5(a) and (b).

The functional forms used so far are the simplest that permit the bifurcations in question. Those bifurcations, however, may occur in systems of any dimension greater than or equal to one. It may be useful to illustrate an example of a (supercritical) pitchfork bifurcation occurring in the two-dimensional system

$$\dot{x} = \mu x - x^3$$
$$\dot{y} = -y. \tag{5.13}$$

In figure 5.6 the bifurcation diagram is illustrated in the space (x, y, μ) indicating: a critical value at $\mu_c = 0$; a single stable node for $\mu < 0$ at $\bar{x} = 0$; the coalescence of three fixed points at $\mu = 0$; two stable nodes and

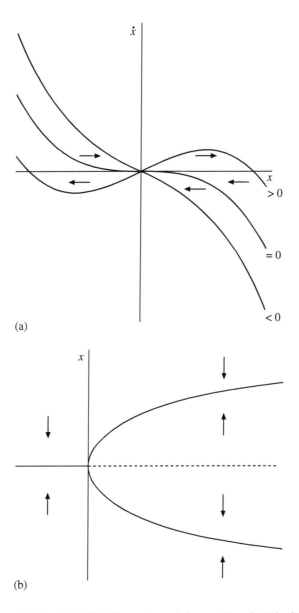

(a)

(b)

Fig. 5.5 Supercritical pitchfork bifurcation: (a) equation (5.11); (b) bifurcation diagram

the saddle for $\mu > 0$. To which of the stable nodes an orbit is eventually attracted depends on the initial value of x.

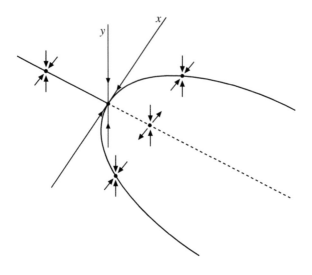

Fig. 5.6 Pitchfork bifurcation diagram for system (5.13)

5.3.4 Generic bifurcations

Notice that, whereas the fold bifurcation is generic, transcritical and pitch-fork bifurcations are not, in the sense that if, starting from a value of μ for which those bifurcations occur, we slightly perturb systems (5.10) and (5.11) (or (5.12)), the family of the perturbed systems will not have a trans-critical or pitchfork bifurcation, although it may have a fold. To see this, consider first the family of differential equations

$$\dot{x} = \mu x - x^2 + \epsilon \qquad \epsilon > 0. \tag{5.14}$$

For $\epsilon = 0$ we are back to (5.10), but for any $\epsilon > 0$ the situation is qualita-tively different. First of all, $x = 0$ is no longer a fixed point. Equilibria are determined by the solutions of the equation

$$x^2 - \mu x - \epsilon = 0$$

namely

$$\bar{x}_{1,2} = \frac{1}{2}\left(\mu \pm \sqrt{\mu^2 + 4\epsilon}\right)$$

and the product $\bar{x}_1 \bar{x}_2 = -\epsilon < 0$. Hence, we have two real solutions of opposite sign for any $\mu \in \mathbb{R}$. For $\mu = 0$, in particular, we have $\bar{x}_1 = +\sqrt{\epsilon}$, $\bar{x}_2 = -\sqrt{\epsilon}$. Considering that $\dot{x} \gtrless 0$ if $x^2 - \mu x - \epsilon \lessgtr 0$, we conclude that locally the positive equilibria are always stable and the negative equilibria

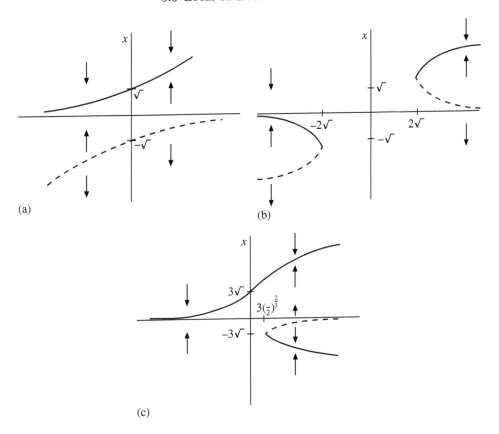

Fig. 5.7 Perturbation away from bifurcation: (a) $+\epsilon$ from transcritical; (b) $-\epsilon$ from transcritical; (c) away from pitchfork

are always unstable. There is no sign of bifurcation for the family of systems in (5.14) as can be observed in the (μ, x) diagram, figure 5.7(a).

If, on the other hand, we add a term $-\epsilon$ to (5.10), the perturbed system is

$$\dot{x} = \mu x - x^2 - \epsilon \qquad \epsilon > 0 \qquad (5.15)$$

and equilibria are solutions to the equation

$$x^2 - \mu x + \epsilon = 0$$

which are

$$\bar{x}_{1,2} = \frac{1}{2}\left(\mu \pm \sqrt{\mu^2 - 4\epsilon}\right).$$

Consequently, for $|\mu| > 2\sqrt{\epsilon}$ there are two positive equilibria if $\mu > 0$, two negative equilibria if $\mu < 0$. In each of these cases the larger equilibria (in algebraic terms) is stable, the smaller unstable. For $|\mu| < 2\sqrt{\epsilon}$, there are no equilibria for (5.15). Thus, $\mu = +2\sqrt{\epsilon}$ and $\mu = -2\sqrt{\epsilon}$ are bifurcation points and in both cases, discontinuous fold bifurcations occur as is represented in figure 5.7(b).

Let us now perturb (5.11)

$$\dot{x} = f(x; \mu) = \mu x - x^3 + \epsilon \qquad \epsilon > 0 \qquad (5.16)$$

for which $\bar{x}_1 = 0$ is no longer an equilibrium. Equilibria of (5.16) are given instead by solutions of the cubic equation

$$x^3 - \mu x - \epsilon = 0. \qquad (5.17)$$

Rather than study (5.17) directly, it is easier to consider the derived function

$$\mu = F(x) = x^2 - \frac{\epsilon}{x} \qquad (5.18)$$

and construct the bifurcation diagram accordingly. There are two separate branches associated with (5.18), according to whether $x \gtrless 0$. If $x > 0$, $F'(x) = 2x + \epsilon/x^2 > 0$, $F''(x) = 2 - 2\epsilon/x^3 \gtrless 0$ if $x \gtrless \sqrt[3]{\epsilon}$. Also $F^{-1}(0) = \sqrt[3]{\epsilon}$; $\lim_{x\to\pm\infty} F(x) = \infty$. If $x < 0$, $F(x) \neq 0$, $F'(x) \gtrless 0$ if $x \gtrless \sqrt[3]{-\epsilon/2}$. At the minimum of the parabola (at which $F'(x) = 0$), we have $\mu = 3(\epsilon/2)^{2/3}$. Finally $\lim_{x\to 0+} F(x) = -\infty$ and $\lim_{x\to 0-} F(x) = +\infty$. Plotting this information in the (μ, x) plane we have the bifurcation diagram as depicted in figure 5.7(c).

For all values of $\mu \in \mathbb{R}$, there exists a positive equilibrium. For $\mu > \mu_c = 3(\epsilon/2)^{2/3}$, there exists, in addition, two (negative) equilibria. When μ decreases through μ_c, these equilibria disappear (and vice versa, when μ increases through μ_c). Thus, at μ_c we have a fold bifurcation. There is no pitchfork bifurcation for the perturbed system (5.16). Considering that $\dot{x} \gtrless 0$ if $\mu x - x^3 + \epsilon \gtrless 0$, we also conclude that the positive equilibria is always stable; the smaller (in algebraic value) of the two negative equilibria is stable and the larger is unstable.

5.3.5 Hopf bifurcation for flows

The Hopf bifurcation is one of the two generic ways in which equilibria of systems of differential equations can lose their stability when one control parameter is changed (the other being the fold bifurcation). In this case, local stability is lost when, changing the value of the controlling parameter, the real part of a pair of complex conjugate eigenvalues of the Jacobian matrix calculated at the equilibrium goes through zero and becomes positive. Near

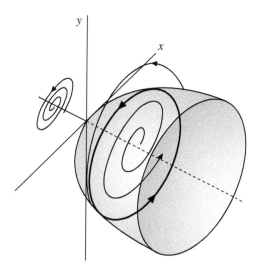

Fig. 5.8 Hopf bifurcation, supercritical

the bifurcation value of the parameter, periodic solutions occur. Of course, the simplest family of systems of differential equations characterised by a Hopf bifurcation must have at least two state variables. The name of the bifurcation derives from a celebrated theorem by Hopf, which generalises certain earlier results of Poincaré and Andronov (and is also known as the Poincaré–Andronov–Hopf theorem).[1] We state the theorem in great generality, but do not discuss its numerous technical details for which we refer the interested reader to the specialised literature (for example, Hassard *et al.*, 1980). The theorem provides a rigorous characterisation of the dynamics of the system near the bifurcation point and states the conditions for stability of the periodic orbits. There exist several, more or less equivalent versions of this theorem, the following is from Invernizzi and Medio (1991).

theorem 5.2 (Hopf) *Consider an autonomous differential equation*

$$\dot{x} = f(x; \mu) \qquad x \in U \qquad \mu \in \mathbb{R} \tag{5.19}$$

where U is an open subset of $\mathbb{R}^m$ and μ is a real scalar parameter varying in some open interval $I \subseteq \mathbb{R}$. Suppose that f is a C^5-function $U \times I \to \mathbb{R}^m$, and that for each μ in I there is an isolated equilibrium point $\bar{x} = \bar{x}(\mu)$. Assume that the Jacobian matrix of f with respect to x, evaluated at $(\bar{x}(\mu); \mu)$, has a pair of simple complex conjugate eigenvalues $\lambda(\mu)$ and $\bar{\lambda}(\mu)$ such that, at

---

[1] For a historical note on the origin of the theorem see Arnold (1980).

a critical value μ_c of the parameter, we have

$$\operatorname{Re} \lambda(\mu_c) = 0 \qquad \operatorname{Im} \lambda(\mu_c) \neq 0 \qquad \frac{d}{d\mu}\operatorname{Re} \lambda(\mu_c) \neq 0,$$

while $\operatorname{Re} \rho(\mu_c) < 0$ *for any other eigenvalue* ρ. *Then (5.19) has a family of periodic solutions. More precisely:*

(i) *there exists a number $\epsilon_0 > 0$ and a C^3-function $r\colon (0, \epsilon_0) \to \mathbb{R}$, with a finite expansion*

$$\mu(\epsilon) = \mu_c + \alpha_2\epsilon^2 + O(\epsilon^4),$$

such that there is a periodic solution $x_\epsilon = x_\epsilon(t)$ of

$$\dot{x} = f[x; \mu(\epsilon)]. \tag{5.20}$$

The coefficient α_2 can be computed by means of certain well-known algorithms.

If $\alpha_2 \neq 0$, then

(ii) *there exists a number ϵ_1, $0 < \epsilon_1 \leq \epsilon_0$, such that $\mu = \mu(\epsilon)$ is bijective from $(0, \epsilon_1)$ either onto $J = (\mu_c, \mu(\epsilon_1))$ when $\alpha_2 > 0$, or onto $J = (\mu(\epsilon_1), \mu_c)$ when $\alpha_2 < 0$. Therefore, the family $x_\epsilon = x_\epsilon(t)$(for $0 < \epsilon < \epsilon_1$) of periodic solutions of (5.20) can be reparametrised by the original parameter μ as $x = x(t; \mu)$.*

Moreover,

(iii) *the stability of $x = x(t; \mu)$ depends on a single Floquet exponent (see chapter 4, section 4.4) $\beta = \beta(\mu)$, which is a C^2-function of $|\mu - \mu_c|^{1/2}$, and it has a finite expansion*

$$\beta = \beta_2\epsilon^2 + O(\epsilon^4)$$

with $\beta_2 = -2\alpha_2(d/d\mu)\operatorname{Re} \lambda(\mu_c)$, and $\epsilon^2 = (\mu - \mu_c)/\alpha_2 + O(\mu - \mu_c)^2$. For μ near μ_c, $x(t; \mu)$ is orbitally asymptotically stable if $\beta_2 < 0$, but it is unstable when $\beta_2 > 0$. Observe that $\beta_2 \neq 0$.

Finally,

(iv) *the period $T = T(\mu)$ of $x(t; \mu)$ is a C^2-function of $|\mu - \mu_c|^{1/2}$, with a finite expansion*

$$T = \frac{2\pi}{|\operatorname{Im} \lambda(\mu_c)|}(1 + \tau_2\epsilon^2 + O(\epsilon^4)),$$

where ϵ^2 is as in (iii) and τ_2 is also a computable coefficient.

For $\beta_2 < 0$, the Hopf bifurcation is called **supercritical** and is of a continuous type, as in figure 5.8. For $\beta_2 > 0$, the bifurcation is called **subcritical** and is of a catastrophic type. In the latter case the limit cycles are not

observable and they can be interpreted as *stability thresholds* of the equilibrium point. Notice that the Hopf bifurcation theorem is *local* in character
and makes predictions only for regions of the parameter and state space of
unspecified size. Unfortunately, the sign of the coefficient β_2 (the **curva-
ture coefficient**), on which stability of the limit cycle depends, is rather
hard to check in any system of dimension larger than two. It requires heavy
calculations involving second and third partial derivatives. This can be a
problem in applications where typically only certain general properties of
the relevant functions are known (e.g., that the relevant function is concave).

5.4 Local bifurcations for maps

Local bifurcation theory for maps generally parallels the theory for flows. To
avoid overburdening the presentation we shall, therefore, discuss the main
points fairly briefly, concentrating our attention on the difference between
continuous- and discrete-time cases.

We are looking for situations in which a fixed point of the map ceases to
be hyperbolic for a certain critical value of the parameter. This happens
typically in three cases: (i) the Jacobian matrix calculated at the fixed point
has one real eigenvalue equal to one; (ii) the eigenvalue is equal to minus one;
(iii) a pair of complex conjugate eigenvalues have modulus equal to one. We
can once more use the centre manifold theorem to reduce the dimensionality
of the problem. The analysis can be reduced to a one-dimensional map in
cases (i) and (ii), to a two-dimensional map in case (iii). Once again the
one-parameter family of maps $G(x; \mu)$ will be treated as a function of two
variables.

5.4.1 Fold, transcritical and pitchfork bifurcations

In the case of an eigenvalue equal to one, three types of local bifurcations
of a fixed point can occur: fold, transcritical and pitchfork (supercritical or
subcritical). For a general one-dimensional family of maps depending on
one parameter,

$$x_{n+1} = G(x_n; \mu) \qquad x_n \in \mathbb{R} \quad \mu \in \mathbb{R}$$

one or other of these local bifurcations occurs if there exists a value μ_c of
the controlling parameter and a corresponding equilibrium value $\bar{x}(\mu_c)$ such

that three of the following conditions are simultaneously satisfied

(i)
$$\frac{\partial G(\bar{x}; \mu_c)}{\partial x_n} = 1$$

for fold, transcritical and pitchfork;

(ii)
$$\frac{\partial^2 G(\bar{x}; \mu_c)}{\partial x_n^2} \neq 0$$

for fold and transcritical;

(ii')
$$\frac{\partial^2 G(\bar{x}; \mu_c)}{\partial x_n^2} = 0 \quad \text{and} \quad \frac{\partial^3 G(\bar{x}; \mu_c)}{\partial x_n^3} \neq 0$$

for pitchfork;

(iii)
$$\frac{\partial G(\bar{x}; \mu_c)}{\partial \mu} \neq 0$$

for fold;

(iii')
$$\frac{\partial G(\bar{x}; \mu_c)}{\partial \mu} = 0 \quad \text{and} \quad \frac{\partial^2 G(\bar{x}; \mu_c)}{\partial \mu \partial x_n} \neq 0$$

for transcritical and pitchfork.

The simplest functional forms for which these bifurcations take place are

$$x_{n+1} = G(x_n; \mu) = \mu - x_n^2 \qquad \text{fold} \tag{5.21}$$
$$x_{n+1} = G(x_n; \mu) = \mu x_n - x_n^2 \quad \text{transcritical} \tag{5.22}$$
$$x_{n+1} = G(x_n; \mu) = \mu x_n - x_n^3 \quad \text{pitchfork.} \tag{5.23}$$

Consider first the fold bifurcation. In figure 5.9(a), curves for (5.21) in the (x_n, x_{n+1}) plane are drawn for three values $\mu > 0$, $\mu = -1/4$, $\mu < -1/4$. From (5.21), putting $x_{n+1} = x_n = \bar{x}$, we find the equilibrium solutions as

$$\bar{x}_{1,2} = \frac{1}{2}(-1 \pm \sqrt{1 + 4\mu}). \tag{5.24}$$

Thus, for $\mu > -1/4$ we have two equilibria. They are nonzero and of opposite signs for $\mu > 0$, both negative for $\mu < 0$, one zero and one negative for $\mu = 0$. There are no real solutions for (5.24), and therefore no equilibria for (5.21), if $\mu < -1/4$. At $\mu = -1/4$ the two equilibria coalesce ($\bar{x}_{1,2} = -1/2$) and, when μ decreases further, they disappear. This information is collected in figure 5.9(b), the bifurcation diagram in the (μ, x) plane where the qualitative change in the behaviour of the system is obvious. At $\bar{x} = -1/2$ and $\mu = -1/4$, $\partial G/\partial x_n = 1$ and the equilibrium is not hyperbolic. Furthermore, $\partial^2 G/\partial x_n^2 = -2$ and $\partial G/\partial \mu = 1$. Then at $\mu =$

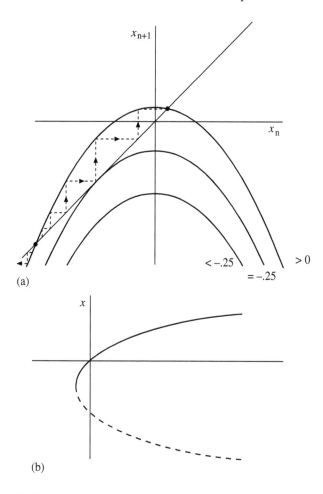

x_{n+1}

x_n

$< -.25$

> 0

$= -.25$

(a)

x

(b)

Fig. 5.9 Fold bifurcation: (a) equation (5.21); (b) bifurcation diagram

$-1/4$ system (5.21) satisfies conditions (i), (ii) and (iii) and there exists a fold bifurcation. Notice that the bifurcation diagram is qualitatively the same as for the flow version of the fold bifurcation, except it has been relocated in the plane.

System (5.22) has two equilibria $\bar{x}_1 = 0$ and $\bar{x}_2 = \mu - 1$. Because $|\partial G/\partial x_n| = |\mu - 2x_n|$, for $-1 < \mu < 1$, $\bar{x}_1 = 0$ is stable and $\bar{x}_2 < 0$ is unstable. Instead, for $1 < \mu < 3$, $\bar{x}_1$ is unstable and $\bar{x}_2 > 0$ is stable. At $\mu = 1$, $\bar{x}_1 = \bar{x}_2 = 0$, $\partial G/\partial x_n = 1$ and $\partial^2 G/\partial x_n^2 = -2$. On the other hand $\partial G/\partial \mu = x_n = 0$ but $\partial^2 G/\partial \mu \partial x_n = 1$. Thus, at $\mu = 1$ conditions (i), (ii), (iii') are satisfied and there exists a transcritical bifurcation, as depicted in figure 5.10(a).

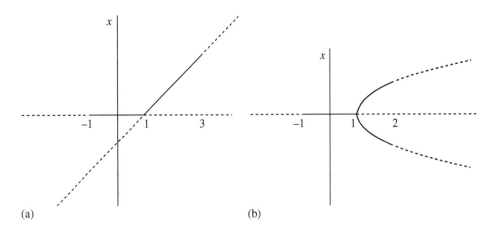

Fig. 5.10 Bifurcations for maps: (a) transcritical; (b) pitchfork

Finally, system (5.23) has an equilibrium $\bar{x}_1 = 0$ for any real value of μ. For $\mu > 1$ it also has two additional equilibria $\bar{x}_{2,3} = \pm\sqrt{\mu - 1}$. Because $|\partial G/\partial x_n| = |\mu - 3x_n^2|$ the null equilibrium is stable for $-1 < \mu < 1$. Both the equilibria $\bar{x}_{2,3}$ are stable at their appearance and over $1 < \mu < 2$. At $\mu = 1$ the three equilibria coalesce to zero and we have $\partial G/\partial x_n = 1$; $\partial^2 G/\partial x_n^2 = -6\bar{x} = 0$ but $\partial^3 G/\partial x_n^3 = -6$; $\partial G/\partial\mu = 0$ but $\partial^2 G/\partial\mu\partial x_n = 1$. Thus, at $\mu = 1$ conditions (i), (ii') and (iii') are satisfied and there exists a (supercritical) pitchfork bifurcation as represented in figure 5.10(b). The simplest functional form leading to a subcritical pitchfork bifurcation is that given by (5.23) if we change the sign of the cubic term.

5.4.2 Flip bifurcation

As we have seen in the last subsection, the bifurcations of one-parameter families of maps occurring when a real eigenvalue of the Jacobian matrix, calculated at equilibrium, goes through plus one are analogous to the corresponding bifurcations for flows when a real eigenvalue goes through zero. However, stability of an equilibrium point of a map can also be lost when a real eigenvalue goes through *minus* one. This leads to a new phenomenon which we investigate by studying the family of maps

$$x_{n+1} = G(x_n; \mu) = \mu x_n(1 - x_n) \qquad x, \mu \in \mathbb{R} \qquad (5.25)$$

where G is the often-mentioned **logistic map**. The equilibrium solutions of (5.25) are $\bar{x}_1 = 0$ and $\bar{x}_2 = 1 - (1/\mu)$. Considering that $\partial G/\partial x_n = \mu(1 - 2x_n)$, we have $\bar{x}_1$ stable for $-1 < \mu < 1$, $\bar{x}_2$ stable for $1 < \mu < 3$. It is

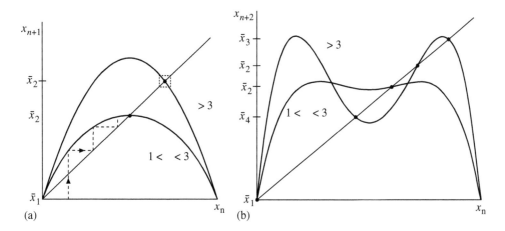

Fig. 5.11 Logistic for 2 values of μ: (a) G; (b) G^2

easy to verify that at $\mu = 1$, conditions (i), (ii) and (iii') of section 5.3 are satisfied and there exists a transcritical bifurcation.

At $\mu = 3$, $\bar{x}_2 = 2/3$, the eigenvalue is equal to minus one and $\bar{x}_2$ is nonhyperbolic. Past $\mu = 3$, both equilibria are unstable. Thus at $\mu = 3$ we have a qualitative change in the behaviour of system (5.25).

To understand this bifurcation better, consider the equation obtained by linearising (5.25) around equilibrium $\bar{x}_2$ for $\mu = 3$, that is,

$$\xi_{n+1} = -\xi_n \qquad (5.26)$$

where $\xi_n \equiv x_n - \bar{x}_2$ for all n. Solution of equations of this type were mentioned in chapter 1, section 1.2, where we discussed improper oscillations. Starting from an initial value ξ_0, an orbit of (5.26) would be $\{\xi_0, -\xi_0, \xi_0, -\xi_0, \ldots\}$, that is, the alternation of two values of the variable, or a cycle of period two. Because the eigenvalue of (5.26) is equal to one in modulus, we cannot invoke the Hartman–Grobman theorem and take solutions of (5.26) as a 'correct' representation of local behaviour of the nonlinear system (5.25). However, further investigation suggests that map (5.25) has indeed, a period-2 cycle for μ slightly greater than three, as depicted in figure 5.11(a), or, equivalently, that its second iterate G^2 has two fixed points (besides $\bar{x}_1$ and $\bar{x}_2$), as depicted in figure 5.11(b).

To determine $\bar{x}_3$ and $\bar{x}_4$ of G^2 set

$$\bar{x} = G[G(\bar{x}; \mu)] = G^2(\bar{x}; \mu)$$

whence

$$\mu^3 \bar{x}^4 - 2\mu^3 \bar{x}^3 + \mu^2(1+\mu)\bar{x}^2 + (1-\mu^2)\bar{x} = 0. \tag{5.27}$$

Equation (5.27) has solutions $\bar{x}_1 = 0$, $\bar{x}_2 = 1 - (1/\mu)$, and two additional solutions

$$\bar{x}_{3,4} = \frac{1 + \mu \pm \sqrt{\mu^2 - 2\mu - 3}}{2\mu} \tag{5.28}$$

where the discriminant is nonnegative for $\mu \geq 3$, and $\bar{x}_2 = \bar{x}_3 = \bar{x}_4 = 2/3$ at $\mu = 3$. Local stability of the equilibria $\bar{x}_3$ and $\bar{x}_4$ of the map G^2 (and thereby local stability of the period-2 cycle of the map G) can be ascertained by evaluating $\partial G^2 / \partial x_n$ calculated at these equilibria. First, notice that $G(\bar{x}_3; \mu) = \bar{x}_4$ and $G(\bar{x}_4; \mu) = \bar{x}_3$, then

$$\frac{\partial G^2(\bar{x}_3; \mu)}{\partial x_n} = \frac{\partial G^2(\bar{x}_4; \mu)}{\partial x_n} = \frac{\partial G(\bar{x}_4; \mu)}{\partial x_n} \frac{\partial G(\bar{x}_3; \mu)}{\partial x_n}.$$

Recalling that $\partial G(x; \mu)/\partial x_n = \mu(1 - 2x)$ we can write

$$\frac{\partial G^2(\bar{x}_3; \mu)}{\partial x_n} = \mu(1 - 2\bar{x}_3)\mu(1 - 2\bar{x}_4) = \mu^2[1 + 4\bar{x}_3\bar{x}_4 - 2(\bar{x}_3 + \bar{x}_4)]$$

and, making use of (5.28), we have

$$\kappa(\bar{x}_3) \equiv \frac{\partial G^2(\bar{x}_3; \mu)}{\partial x_n} = -\mu^2 + 2\mu + 4$$

where $\kappa(\bar{x}_3)$ is the eigenvalue of interest and $\kappa(\bar{x}_4) = \kappa(\bar{x}_3)$. At $\mu = 3$, $\kappa(\bar{x}_3) = \kappa(\bar{x}_4) = 1$ and for μ slightly larger than three, these eigenvalues are smaller than one. Therefore, the two fixed points $\bar{x}_3$ and $\bar{x}_4$ of the map G^2 are initially (for μ slightly larger than three) stable, as is the period-2 cycle of the map G.

The bifurcation of a fixed point of G occurring when its eigenvalue passes through minus one, the nonzero fixed point loses its stability and a stable period-2 cycle is born, is called a **flip** bifurcation.

remark 5.4 To the flip bifurcation for the map G there corresponds a pitchfork bifurcation for the map G^2. In order to avoid any confusion, it should always be specified to which map the bifurcation refers.

For a general one-dimensional family of maps $G(x_n; \mu)$, the necessary and sufficient conditions for a flip bifurcation to occur at $\bar{x}(\mu_c)$ are

(i') $$\frac{\partial G(\bar{x}; \mu_c)}{\partial x_n} = -1$$

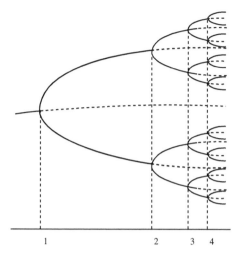

Fig. 5.12 Sequence of flip bifurcations

(ii'') $\qquad \dfrac{\partial^2 G^2(\bar{x}; \mu_c)}{\partial x_n^2} = 0 \quad \text{and} \quad \dfrac{\partial^3 G^2(\bar{x}; \mu_c)}{\partial x_n^3} \neq 0$

(iii'') $\qquad \dfrac{\partial G^2(\bar{x}; \mu_c)}{\partial \mu} = 0 \quad \text{and} \quad \dfrac{\partial^2 G^2(\bar{x}; \mu_c)}{\partial \mu \partial x_n} \neq 0$

where $\bar{x}$ is the bifurcating fixed point and μ_c the critical or bifurcation value of the controlling parameter. The flip bifurcation of G and the related pitchfork bifurcation of G^2 are supercritical or subcritical according to whether

$$-\frac{\partial^3 G^2(\bar{x}; \mu_c)}{\partial x_n^3} \bigg/ \frac{\partial^2 G^2(\bar{x}; \mu_c)}{\partial \mu \partial x_n} \gtrless 0$$

respectively.

Now let us return to the logistic map (5.25). If μ is increased beyond the (flip) bifurcating value, the period-2 cycle of the map G (as well as the equilibria of G^2, $\bar{x}_3$ and $\bar{x}_4$) will lose stability at $\mu = 1 + \sqrt{6}$, at which $\kappa(\bar{x}_3) = \kappa(\bar{x}_4) = -1$. At that value of μ, a new flip bifurcation occurs, this time for the map G^2, with loss of stability of equilibria $\bar{x}_3$ and $\bar{x}_4$. Consequently, an initially stable period-2 cycle for G^2, and therefore a period-4 cycle for G, is born. The period-doubling scenario continues as we increase μ further. A schematic representation of this sequence of flip bifurcations is illustrated in figure 5.12. We will provide a more detailed analysis of this scenario in chapter 8, when we discuss routes to chaos.

5.4.3 Neimark–Sacker bifurcation

The case in which a pair of complex eigenvalues of the Jacobian matrix at a fixed point of a map has modulus one is analogous, but not quite the same as the corresponding case for flows. The bifurcation occurring in this case is often referred to as 'Hopf bifurcation for maps' but, as a matter of fact, the relevant results were first stated by Neimark (1959) and subsequently proved by Sacker (1965).[2]

theorem 5.3 (Neimark–Sacker) *Let $G_\mu : \mathbb{R}^2 \to \mathbb{R}^2$ be a family of maps of the class C^k, $k \geq 5$, depending on a real parameter μ, so that for μ near 0, $\bar{x} = 0$ is a fixed point of G_μ and the following conditions are satisfied*

- (i) *for μ near zero, the Jacobian matrix has two complex, conjugate eigenvalues $\kappa(\mu)$ and $\bar{\kappa}(\mu)$ with $|\kappa(0)| = 1$;*
- (ii) $\frac{d|\kappa(0)|}{d\mu} \neq 0$;
- (iii) $[\kappa(0)]^i \neq 1$, *for $i = 1, 2, 3$ and 4, that is, $\kappa(0)$ is not a low root of unity.*

Then, after a trivial change of the μ coordinate and a smooth, μ-dependent coordinate change on $\mathbb{R}^2$,

- (i) *the map G_μ in polar coordinates takes the form:*

$$\begin{pmatrix} r_{n+1} \\ \phi_{n+1} \end{pmatrix} = \begin{pmatrix} (1+\mu)r_n - \alpha(\mu)r_n^3 \\ \phi_n + \beta(\mu) + \gamma(\mu)r_n^2 \end{pmatrix} + O\left(\left|\begin{matrix} r_n \\ \phi_n \end{matrix}\right|^5\right) \qquad (5.29)$$

  where α, β, γ are smooth functions of μ and $\alpha(0) \neq 0$;
- (ii) *for $\alpha > 0$ (respectively, for $\alpha < 0$) and in a sufficiently small right (left) neighbourhood of $\mu = 0$, for the map G_μ there exists an invariant attractive (repelling) circle Γ_μ bifurcating from the fixed point at $\bar{x} = 0$ and enclosing it.*

Assuming now that $\alpha > 0$ and the invariant circle is (locally) attractive, the behaviour of G_μ restricted to the set Γ_μ can be approximated arbitrarily well by iterations of a homeomorphism of the unit circle f_μ. The dynamics of the latter depend crucially on an invariant of f_μ called the **rotation number** defined as

$$\rho(f) = \lim_{n \to \infty} \frac{\hat{f}^n(x) - x}{n},$$

where the index μ is omitted and the map $\hat{f}$ is the so-called **lift** of f, that is, a map $\hat{f} : \mathbb{R} \to \mathbb{R}$ such that, setting $h : \mathbb{R} \to S^1$, $h(x) = e^{i2\pi x}$, we have

---

[2]For a detailed discussion of this bifurcation, see Iooss (1979) and Whitley (1983).

$h \circ \hat{f} = f \circ h$. The lift of f is unique up to the addition of an integer. The limit $\rho(f)$ exists and is independent of x and the choice of $\hat{f}$.

There are two basic possibilities:

(i) The limit $\rho(f) = p/q$, where p and q are two integers, that is, ρ is rational. In this case, the map f has periodic orbits of period q.

(ii) The limit $\rho(f)$ is irrational. In this case, a known result[3] states that if $f \in C^2$, then it is topologically conjugate (an equivalence relation that we study in detail in chapter 6) and therefore dynamically equivalent to a fixed rotation of the circle; that is, there exists a homeomorphism θ such that $\theta \circ f = G_\rho \circ \theta$, where, using the additive notation,

$$G_\rho(x) = x + \rho \quad \mod 1 \tag{5.30}$$

and ρ is the rotation number.

For ρ irrational, (5.30) is the prototypical system generating quasiperiodic orbits which we discussed in chapter 4, section 4.5. We shall return to the issue in chapter 9 where we deal with the ergodic theory of dynamical systems. Notice that in the approximation given by (5.29), the dynamics of the system are decoupled in the sense that the first equation determines the stability of the cycle, while the second equation describes the dynamics *on* the circle. Ignoring the higher-order terms in (5.29), we can verify that the radius of the invariant circle is equal to $r = \sqrt{\mu/\alpha(\mu)}$. The similarities between the Hopf bifurcation for flows and the Neimark–Sacker bifurcation for maps are obvious but an important difference should be emphasised. Whereas the invariant limit cycle originating from the former consists of a single orbit, Γ_μ is an invariant set on which there exist *many different* orbits.

remark 5.5 When condition (iii) in theorem 5.3 is violated, that is, an eigenvalue is equal to one of the first four roots of unity, the situation is much more complicated and is referred to as 'strong resonance'. The cases $\kappa^i = 1$, $(i = 1, 2)$ occur when there exist double eigenvalues equal to one or negative one, respectively. The corresponding bifurcations are of codimension-2 (i.e., they depend on two parameters). The more interesting cases κ^i $(i = 3, 4)$ have been studied by Arnold (1980).

---

[3]This result is known as the Denjoy theorem. For a discussion, proofs and bibliography, see Katok and Hasselblatt (1995), chapter 12, pp. 401 ff.

5.5 Bifurcations in two-dimensional maps

For two-dimensional maps the search for nonhyperbolic fixed points and the determination of the way in which the stability frontier is crossed is simplified by the use of stability conditions on the coefficients of the characteristic equation. Recall from chapter 3 that a fixed point $\bar{x}$ is (locally) asymptotically stable if the eigenvalues κ_1 and κ_2 of the Jacobian matrix, calculated at the fixed point, are less than one in modulus. The necessary and sufficient conditions guaranteeing that $|\kappa_1| < 1$ and $|\kappa_2| < 1$ are:

(i) $$1 + \operatorname{tr} DG(\bar{x}) + \det DG(\bar{x}) > 0$$

(ii) $$1 - \operatorname{tr} DG(\bar{x}) + \det DG(\bar{x}) > 0$$

(iii) $$1 - \det DG(\bar{x}) > 0.$$

Each of the three generic bifurcations results from the loss of stability through the violation of one of these three conditions (refer to figure 2.7):

(i) The flip bifurcation occurs when a single eigenvalue becomes equal to -1, that is, $1 + \operatorname{tr} DG(\bar{x}) + \det DG(\bar{x}) = 0$, with $\operatorname{tr} DG(\bar{x}) \in (0, -2)$, $\det DG(\bar{x}) \in (-1, 1)$ (i.e., conditions (ii) and (iii) are simultaneously satisfied).

(ii) The fold bifurcation occurs when a single eigenvalue becomes equal to $+1$, that is, $1 - \operatorname{tr} DG(\bar{x}) + \det DG(\bar{x}) = 0$, with $\operatorname{tr} DG(\bar{x}) \in (0, 2)$, $\det DG(\bar{x}) \in (-1, 1)$ (i.e., conditions (i) and (iii) are simultaneously satisfied). For particular cases, the violation of (ii) might lead to a transcritical or pitchfork bifurcation.

(iii) The Neimark–Sacker bifurcation occurs when the modulus of a pair of complex, conjugate eigenvalues is equal to one. Since the modulus of complex eigenvalues in $\mathbb{R}^2$ is simply the determinant of $DG(\bar{x})$, this occurs at $\det DG(\bar{x}) = 1$. If, moreover, conditions (i) and (ii) are simultaneously satisfied (i.e., $\operatorname{tr} DG(\bar{x}) \in (-2, 2)$), there may be a Neimark bifurcation.

remark 5.6 These are, of course, necessary conditions for the existence of the associated bifurcation. However, combined with numerical simulations suggesting that such bifurcations do occur, they constitute strong evidence. In applications where analytical proofs can be difficult to provide, this procedure is, at times, the only available.

This concludes our introductory study of bifurcations. In the following chapters we see how bifurcation theory is related to the transition of systems to chaos.

Exercises

5.1 Consider the system $\dot{x} = f(x)$ with a hyperbolic fixed point at $\bar{x} = 0$. Perturb it slightly, that is, consider the perturbed system $\dot{x} = f(x) + \epsilon g(x)$. Show that, if ϵ is sufficiently small, the perturbed system will have a hyperbolic fixed point near zero, and the Jacobian matrices of the original and the perturbed system have the same eigenvalue structure.

5.2 For (5.10), $\dot{x} = \mu x - x^2$, prove that since conditions (ii) and (iii') are satisfied, besides $x = 0$ there exists a second curve of fixed points $\bar{x}(\mu)$ not coinciding with $x = 0$.

5.3 For the following equations: sketch the function $\dot{x}$ against x for $\mu = 0$, $\mu > 0$, and $\mu < 0$; indicate fixed points; use arrows to suggest their stability; translate the information from $(x, \dot{x})$ to the bifurcation diagram in (μ, x)

(a) $\dot{x} = \mu - \frac{x^2}{4}$
(b) $\dot{x} = -\mu x + x^3$
(c) $\dot{x} = \mu - |x|$
(d) $\dot{x} = \mu - x^3$
(e) $\dot{x} = \mu - x^2 + 4x^4$.

5.4 For the equations in exercise 5.3 determine the (real) fixed points, the interval of μ over which they exist, the intervals for which they are stable or unstable, the values at which they are nonhyperbolic. Demonstrate the existence of codimension-1 bifurcations and indicate whether they are continuous or discontinuous and, where relevant, whether they are supercritical or subcritical.

5.5 Prove that for (5.15) the larger fixed point of the equilibrium pair

$$\bar{x}_{1,2} = \frac{1}{2}\left(\mu \pm \sqrt{\mu^2 - 4\epsilon}\right)$$

is stable, the smaller of the pair unstable.

5.6 Sketch the bifurcation diagram of

$$\dot{x} = x^3 + x^2 - (2 + \mu)x + \mu.$$

(*Hint*: there is a fixed point at $\bar{x}_1 = 1$.)

5.7 Discuss the conditions for the existence of a Hopf bifurcation for the system:

$$\dot{x} = y + kx(x^2 + y^2)$$
$$\dot{y} = -x + \mu y.$$

5.8   For the following equations: determine the intervals of μ over which
        (real) fixed points exist, the intervals for which they are stable or un-
        stable, the values of μ at which they are nonhyperbolic. Demonstrate
        the existence of codimension-1 bifurcations and indicate whether
        they are continuous or discontinuous and, where relevant, whether
        they are supercritical or subcritical. Sketch the bifurcation diagram

        (a) $x_{n+1} = \mu - x_n^2$
        (b) $x_{n+1} = -\mu x_n + x_n^3$
        (c) $x_{n+1} = \mu + x_n - x_n^2$
        (d) $x_{n+1} = \mu x_n + x_n^3$.

5.9   Sketch the bifurcation diagram and label the bifurcations for the
        system

$$x_{n+1} = \frac{1}{2}\mu x_n - 2x_n^2 + x_n^3.$$

5.10   Verify that conditions (i'), (ii'') and (iii'') are satisfied for the map
        (5.25) at $\bar{x} = 2/3$ and $\mu_c = 3$.

5.11   The following system derives from an economic model of the class
        known as 'overlapping generations' models (for details on this spe-
        cific model see Medio, 1992, chapter 12):

$$c_{n+1} = l_n^\mu \qquad\qquad \mu > 1$$
$$l_{n+1} = b(l_n - c_n) \qquad b > 1.$$

        Determine the nontrivial fixed point of the system. Use the Jacobian
        matrix calculated at that equilibrium and the stability conditions
        given in section 5.5 to determine the **stability frontier**. The latter
        is the curve which separates parameter configurations leading to
        eigenvalues with modulus less than one from those with modulus
        greater than one. Determine also the **flutter boundary**, that is,
        the curve separating real and complex eigenvalues. Sketch these two
        curves in the parameter space, (μ, b) and label the type of dynamics
        in each area. Discuss the bifurcations occurring when the stability
        frontier is crossed.

6

Chaotic sets and chaotic attractors

Invariant, attracting sets and attractors with a structure more complicated than that of periodic or quasiperiodic sets are called chaotic. Before providing precise mathematical definitions of the properties of chaotic systems, let us first try to describe them in a broad, nonrigorous manner. We say that a discrete- or continuous-time dynamical system is chaotic if its typical orbits are aperiodic, bounded and such that nearby orbits separate fast in time. Chaotic orbits never converge to a stable fixed or periodic point, but exhibit sustained instability, while remaining forever in a bounded region of the state space. They are, as it were, *trapped unstable orbits*. To give an idea of these properties, in figure 6.1 we plot a few iterations of a chaotic discrete-time map on the interval, with slightly different initial values. The two trajectories remain close for the first few iterations, after which they separate quickly and have thereafter seemingly uncorrelated evolutions. Time series resulting from numerical simulations of such trajectories *look* random even though they are generated by deterministic systems, that is, systems that do not include any random variables. Also, the statistical analysis of deterministic chaotic series by means of certain *linear* techniques, such as estimated autocorrelation functions and power spectra, yields results similar to those determined for random series. Both cases are characterised by a rapidly decaying autocorrelation function and a broadband power spectrum.

remark 6.1 Notice that one or another (but not all) of the properties listed above can be found in nonchaotic systems as well. For example, a linear system characterised by a unique unstable fixed point will have aperiodic orbits separating exponentially fast but they will not be bounded. On the other hand, quasiperiodic systems will have aperiodic, bounded solutions but not separating orbits.

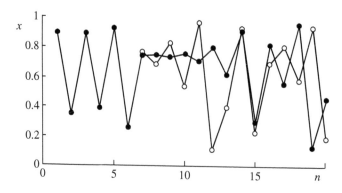

Fig. 6.1 Separation of nearby trajectories, black circles $x_0 = 0.9005$, white circles
$x_0 = 0.9$

remark 6.2 In this book we discuss only *bounded* chaotic orbits and sets.
There exist some recent, interesting results on unbounded chaotic dynamics
to which we refer the reader (see, for example, Bischi *et al.*, 2000).

To prove rigorous and general results for chaotic solutions is very difficult,
especially so with regards to chaotic *attractors*. Therefore, a substantial
part of mathematical (as distinguished from physical) research on chaos has
been concentrated on the study of properties of chaotic sets which are invariant, but not necessarily attracting. In this book we are concerned with
the application of dynamical system theory and our main interest is the observable behaviour of those systems, hence the properties of their attracting
sets. However, the detailed study of some simple models possessing invariant, though not necessarily attracting sets with a complex orbit structure is
the first necessary step towards the understanding of the basic mechanisms
generating chaotic dynamics.

While in this chapter and chapter 7 we concentrate on topological or
geometric aspects of chaos, in chapters 9 and 10 we investigate the ergodic
or probabilistic features of chaotic systems. Most of the examples discussed
below are discrete-time systems (maps) for which precise results are easier
to obtain, but some important continuous-time cases will also be considered.

6.1 Basic definitions

There exist several, partly overlapping characterisations of chaos. We shall
adopt here the following definitions.

definition 6.1 *A flow ϕ (respectively, a continuous map G) on a metric space M is said to possess* **sensitive dependence on initial conditions** *(SDIC) on M if there exists a real number $\delta > 0$ such that for all $x \in M$ and for all $\epsilon > 0$, there exists $y \in M$, $y \neq x$, and $T > 0$ (respectively, an integer $n > 0$) such that $d(x, y) < \epsilon$ and $d[\phi(T, x), \phi(T, y)] > \delta$ (respectively, $d[G^n(x), G^n(y)] > \delta$).*

For example, notice how the trajectories in figure 6.1, which have a negligible initial distance, after few iterations are widely separate.

definition 6.2 *A flow ϕ (respectively, a continuous map G) is said to be* **chaotic** *on a compact invariant set A if:*

(a) it is topologically transitive on A, according to definitions 4.8 or 4.9
(b) it has sensitive dependence on initial conditions on A, according to definition 6.1.

In this case, some authors would say that 'A is chaotic for ϕ (or for G)'.

remark 6.3 Some clarifications are in order.

(1) Condition *(a)* of definition 6.2 is introduced to make sure that we are talking about a single indecomposable invariant set.
(2) Condition *(b)* can be made sharper (and more restrictive) in two ways. First, we may require that divergence of nearby points takes place at an exponential rate. This property can be made more precise by means of the concept of a Lyapunov characteristic exponent (LCE). Broadly speaking, LCEs measure the exponential rate at which nearby orbits diverge in the different directions. (A more detailed discussion of LCEs is given in chapter 7, section 7.1 and chapter 9, section 9.2.) Secondly, we may require that divergence (exponential or otherwise) occurs for *each pair $x, y \in A$*. In this case, the flow ϕ or map G is called **expansive** on A.
(3) In order to appreciate the importance of the requirement that A be compact (which, for $A \subset \mathbb{R}^m$ is equivalent to A being closed and bounded), consider the following case (cf. Wiggins, 1990, p. 609):

$$\dot{x} = ax \qquad x \in \mathbb{R} \quad a > 0. \tag{6.1}$$

Equation (6.1) is linear and its solution is $\phi(t, x) = x_0 e^{at}$. Therefore the flow map ϕ is topologically transitive on the open, unbounded (and therefore noncompact) invariant sets $(-\infty, 0)$ and $(0, \infty)$. Also, for any two points $x_1, x_2 \in \mathbb{R}$ and $x_1 \neq x_2$ we have

$$|\phi(t, x_1) - \phi(t, x_2)| = e^{at}|x_1 - x_2|$$

and ϕ has SDIC on $\mathbb{R}$ (in fact, it is expansive). However, the orbits generated by ϕ are not chaotic.

(4) Definition 6.2 is essentially the same as that of Wiggins (1990), p. 608 or Robinson (1999), p. 86. Notice that this definition refers to a 'chaotic flow (or map) on a set A' or, for short, a 'chaotic set A'. It does not imply that *all orbits* of a chaotic flow (or map) on A are chaotic. In fact, there are many nonchaotic orbits on chaotic sets, in particular, there are many *unstable* periodic orbits. So important are they that certain authors (e.g., Devaney, 1986, p. 50, definition 8.5) add a third condition for chaos, that *periodic orbits are dense* on A. This is an interesting property and it is automatically satisfied if the chaotic invariant set is hyperbolic (cf. Wiggins, 1990, p. 615; Robinson, 1999, p. 86) but we prefer not to include it in a basic definition of chaos.

(5) The close relation between chaos, as characterised in definition 6.2, and dense periodic sets is confirmed by two quite general results. The first, established by Banks *et al.* (1992) states that for any continuous map on a metric space, transitivity and the presence of a dense set of periodic orbits imply sensitive dependence on initial conditions, that is, chaos. The second result, by Vellekoop and Berglund (1994) states that for any continuous map on an interval of $\mathbb{R}$, transitivity alone implies the presence of a dense set of periodic orbits and therefore, in view of the first result, it implies sensitive dependence on initial conditions, and therefore chaos.

(6) Certain authors define chaos in terms of *orbits* rather than sets. For example, Alligood *et al.* (1997) (p. 196, definition 5.2; p. 235, definition 6.2; pp. 385–6, definition 9.6) define a chaotic set as the ω-limit set of a chaotic orbit $\{G^n(x_0)\}$ which itself is contained in its ω-limit set. In this case, the presence of sensitive dependence on initial conditions (or a positive Lyapunov characteristic exponent) is not enough to characterise chaotic properties of orbits and additional conditions must be added to exclude unstable periodic or quasiperiodic orbits.

Other related characterisations of chaos, positive entropy and mixing will be discussed in chapter 9 when the basic concepts of ergodic theory have been developed.

6.2 Symbolic dynamics and the shift map

Equipped with the definitions of section 6.1, we now discuss a technique known as *symbolic dynamics* which provides a very powerful tool for understanding the orbit structure of a large class of dynamical systems. Broadly speaking, the approach can be summarised as follows.

First, we define a certain auxiliary system characterised by a map, called a shift map, acting on a space of infinite sequences called the symbol space. Next, we prove certain properties of the shift map. Finally, we establish a certain equivalence relation between a map we want to study and the shift map, and show that the relation preserves the properties in question.

We begin by defining the symbol space and the shift map. Let S be a collection of symbols. In a physical interpretation, the elements of S could be anything, for example, letters of an alphabet or discrete readings of some measuring device for the observation of a given dynamical system. To crystallise ideas, we assume here that S consists of only two symbols, let them be 0 and 1. Then we have $S = \{0, 1\}$. Next, we want to construct the space of all possible bi-infinite sequences of 0 and 1, defined as

$$\Sigma_2 \equiv \cdots S \times S \times S \times \cdots.$$

A point in Σ_2 is therefore represented as a **bi-infinity-tuple** of elements of S, that is,

$$s \in \Sigma_2$$

means

$$s = \{\ldots s_{-n} \ldots s_{-1} s_0 s_1 \ldots s_n \ldots\}.$$

where $\forall i$, $s_i \in S$ (i.e., $s_i = 0$ or 1). For example

$$s = \{\ldots 0\,0\,1\,0\,1\,1\,1\,0\,1 \ldots\}.$$

We can define a distance function $\bar{d}$ in the space Σ_2

$$\bar{d}(s, \bar{s}) = \sum_{i=-\infty}^{+\infty} \frac{d(s_i, \bar{s}_i)}{2^{|i|}} \tag{6.2}$$

where d is the discrete distance in $S = \{0, 1\}$, that is, $d(s_i, \bar{s}_i) = 0$ if $s_i = \bar{s}_i$ and $d = 1$ if $s_i \neq \bar{s}_i$. This means that two points of Σ_2, that is two infinite sequences of 0s and 1s, are close to one another if their *central elements* are close, i.e., if the elements with a small, positive or negative index are close. Notice that, from the definition of $\bar{d}(s_i, \bar{s}_i)$, the infinite sum on the RHS of (6.2) is less than or equal to 3 and, therefore, converges.

Next, we define the **shift map** on Σ_2 as

$$T : \Sigma_2 \to \Sigma_2 \qquad T(s) = s' \text{ and } s'_i = s_{i+1}$$

that is,

$$s = (\ldots s_{-2} s_{-1} s_0 \; s_1 \; s_2 \ldots)$$

$$T(s) = s' = (\ldots s'_{-2} s'_{-1} \; s'_0 \; s'_1 \; s'_2 \ldots)$$
$$= (\ldots s_{-1} \; s_0 \; s_1 \; s_2 \; s_3 \ldots).$$

In words, the map T takes a sequence and shifts each entry one place to the left. It is sometimes called a **topological Bernoulli shift** for reasons that will become clear in chapter 10.

We can similarly define the **one-sided shift map** T_+ on the space Σ_{2+} of one-sided infinite sequences of two symbols, that is, $s \in \Sigma_{2+}$ means $s = \{s_0 s_1 \ldots s_n \ldots\}$. In this case, we have

$$T_+ : \Sigma_{2+} \to \Sigma_{2+} \qquad T_+(s) = s' \text{ and } s'_i = s_{i+1}$$

so that

$$T_+(s_0 s_1 s_2 \ldots) = (s'_0 s'_1 s'_2 \ldots) = (s_1 s_2 s_3 \ldots).$$

In other words, the T_+ map shifts a one-sided sequence one place to the left and drops its first element. The maps T and T_+ have very similar properties, but whereas T is invertible, T_+ is not.

For the space Σ_{2+}, we can use essentially the same distance function as (6.2) with the obvious difference that the infinite sum will now run from zero to $+\infty$. In this case, two unilateral sequences are close if their initial elements are close. In what follows, we shall mostly consider the one-sided shift map T_+ because it can be used to prove chaotic properties of certain noninvertible, one-dimensional maps frequently employed in applications. Some of the results proved for T_+ can be easily extended to T.

We now establish the following.

proposition 6.1 *The shift map T_+ on Σ_{2+} is chaotic according to definition 6.2.*

We can verify proposition 6.1 in three steps:

(i) Σ_{2+} is compact and (forward) invariant with respect to the shift map T_+. Invariance is obvious: if we apply T_+ to a sequence in Σ_{2+} we obtain another sequence in Σ_{2+}. Compactness of Σ_{2+} is the consequence of the fact that, by construction

$$\Sigma_{2+} \equiv S \times S \times S \cdots$$

that is, Σ_{2+} is a unilaterally infinite Cartesian product of copies of the set $S \equiv \{0, 1\}$. The latter is a discrete space consisting of just two points and is compact. From a known theorem in topology due to Tychonoff (see Dugundji, 1966, p. 224; Wiggins, 1990, p. 442) it follows that Σ_{2+} is compact if S is.

(ii) The map T_+ is topologically transitive on Σ_{2+}. Consider a unilateral sequence s^* that lists first all the blocks consisting of one symbol (in this case, 0 or 1); next, all the blocks consisting of two symbols (in this case, $(00), (01), (10), (11)$); then, all the blocks of three symbols and so on and so forth. The sequence s^* is a point in Σ_{2+}. Now take *any other sequence* (any other point $s \in \Sigma_{2+}$) and any $k \geq 1$. Then there is an integer n such that the nth iteration of T_+ starting from s^* generates a sequence, $t \equiv T_+^n(s^*)$, which agrees with s in the first k places. The distance between t and s will be

$$\bar{d}(t, s) = \sum_{i=0}^{\infty} \frac{d(t_i, s_i)}{2^i} \leq \sum_{i=k+1}^{\infty} 2^{-i}$$

which can be made as small as desired by increasing k. Thus we have shown that the set generated by the forward orbit of s^* under T_+ is dense in Σ_{2+}, and the map, T_+, is topologically transitive on Σ_{2+} according to definition 4.9. The proof could be extended to shift maps on N symbol spaces (cf. Devaney, 1986, p. 42; Katok and Hasselblatt, 1995, pp. 48–9; Robinson, 1999, pp. 40–1).

(iii) The next step is to demonstrate that T_+ has SDIC. Let us choose δ of definition 6.1 equal to one. Then take *any* two points in Σ_{2+} corresponding to two sequences s and t. If $s \neq t$, there must be an n such that $T_+^n(t)$ and $T_+^n(s)$ differ in the first element. Therefore,

$$\bar{d}\left[T_+^n(t), T_+^n(s)\right] \geq 1.$$

Then T_+ is expansive and therefore is characterised by SDIC on Σ_{2+} (cf. Robinson, 1999, p. 87). From (i), (ii) and (iii), it follows that T_+ is chaotic on the compact set Σ_{2+}.

remark 6.4 The shift map T_+ on Σ_{2+} has a stronger form of topological transitivity called topologically mixing. In general, we say that a map G is **topologically mixing** on a set A if for any two open subsets U and V of A there exists a positive integer N such that $G^n(U) \cap V \neq \emptyset$ for all $n \geq N$. If a map G is topologically mixing, then for any integer n the map G^n is topologically transitive.

remark 6.5 The arguments used to verify proposition 6.1 can be used, with slight modifications, to demonstrate that the shift map T is topologically transitive and has SDIC on Σ_2 (the space of bi-infinite sequences of 0 and 1).

The fact that the shift map has been proved to be chaotic in a precise sense is all the more interesting because the chaotic properties of invariant sets of certain one- and two-dimensional maps and three-dimensional flows may sometimes be proved by showing that the dynamics on these sets are *topologically conjugate* to that of a shift map on a symbol space. This indirect argument is often the only available strategy for investigating nonlinear maps (or flows) and we discuss some basic examples in the next sections. Although it has not been explicitly introduced we have already encountered the concept of topological conjugacy in the Hartman–Grobman theorem, theorem 3.1 (which we called homeomorphic equivalence) between a nonlinear map (or flow) and its linearisation in a neighbourhood of a fixed point. We now provide some formal definitions.

definition 6.3 *Let $F : X \to X$ and $G : Y \to Y$ be two continuous maps on two arbitrary metric spaces X and Y. A map $h : X \to Y$ that is continuous, onto (surjective) and for which $h \circ F = G \circ h$ is called a* **topological semiconjugacy** *from F to G. The map h is called a* **topological conjugacy** *if it is a semiconjugacy and h is a homeomorphism (h is also one-to-one with continuous inverse). The maps F and G are then called topologically semiconjugate or conjugate, respectively. In the case of semiconjugacy, we also say that G is a* **topological factor** *of F.*

Two topologically conjugate maps F and G are equivalent in their dynamics. In particular, if $\bar{x}$ is a fixed point of F then $h(\bar{x})$ is a fixed point of G, and $\bar{x}$ and $h(\bar{x})$ have the same stability properties; periodic orbits of F are mapped by h to periodic orbits of G. The image under h of the ω-limit of F is the ω-limit of G; topological transitivity (or even topological mixing) and SDIC (on compact spaces) are preserved by conjugacy. Thus, if F is chaotic so is G (and vice versa). We shall see later that semiconjugacy may also be exploited to establish the chaotic properties of a map.

remark 6.6 Two flows ϕ, ψ of a compact manifold M are said to be **topologically equivalent** if there exists a homeomorphism $h : M \to M$, that takes orbits of ϕ to orbits of ψ preserving their orientation. This means that if $p \in M$ and $\delta > 0$, there exists $\epsilon > 0$ such that for $0 < t < \delta$,

$$h[\phi(t, p)] = \psi[t', h(p)]$$

for some $0 < t' < \epsilon$. The map h is called a topological equivalence between

ϕ and ψ. The two flows are **topologically conjugate** if there exists a topological equivalence h that preserves the parameter t, that is

$$h[\phi(t, p)] = \psi[t, h(p)] \qquad \forall p \in M \quad \forall t \in \mathbb{R}.$$

6.3 Logistic map with $\mu > 2 + \sqrt{5}$

A first example of a chaotic invariant set analysed by means of a conjugacy with the shift map is provided by the logistic map

$$x_{n+1} = G(x_n) = \mu x_n(1 - x_n)$$

with $\mu > 2 + \sqrt{5}$ (the result could be proved for $\mu > 4$ but the proof is more difficult and we omit it here). The following discussion is based on Katok and Hasselblatt (1995), pp. 80–1.

Let us consider the closed interval $I = [0, 1]$. It is easy to see that, starting from most points $x \in I$, the orbits generated by G will leave I (in fact, they will tend to $-\infty$). The set Λ of points of I that never leave that set (assuming that Λ is nonempty) is given by

$$\Lambda = \bigcap_{n=0}^{\infty} G^{-n}([0, 1]).$$

The pre-image of $[0, 1]$ consists of the two nonoverlapping subsets

$$\Delta^L : \left[0, \frac{1}{2} - \sqrt{\frac{1}{4} - \frac{1}{\mu}}\right] \qquad \Delta^R : \left[\frac{1}{2} + \sqrt{\frac{1}{4} - \frac{1}{\mu}}, 1\right].$$

The pre-image of the sets Δ^L and Δ^R consists, respectively, of the (nonoverlapping) subsets

$$G^{-1}(\Delta^L) = \Delta^{LL} \cup \Delta^{RR}$$
$$G^{-1}(\Delta^R) = \Delta^{LR} \cup \Delta^{RL}$$

as illustrated in figure 6.2, and so on.

If we consider that $|dG/dx| > 1$ for $x \in \Delta^L \cup \Delta^R$, then for any sequence $\{w_n\} = (w_0, w_1, \ldots)$ with $w_n \in \{L, R\}$, the length of the subsets defined by $\cap_{n=0}^{j} G^{-n}(\Delta^{w_n})$ decreases exponentially with n and tends to zero as $j \to \infty$. Then Λ is an infinite set of points belonging to the class of Cantor sets.[1]

---

[1] An infinite set of points is called a **Cantor set** if: (i) it is closed; (ii) it is totally disconnected, i.e., it contains no intervals; (iii) it is perfect, i.e., every point in it is an accumulation point or limit point of other points in the set. We shall return to the Cantor set in our discussion of fractal dimension in chapter 7, section 7.2, and of the transition to chaos in chapter 8.

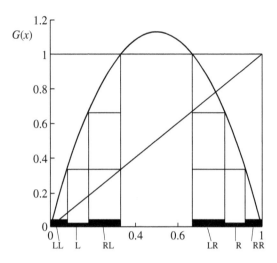

Fig. 6.2 Pre-images of intervals under the map G

Consider now the space Σ_{2+} of one-sided sequences of the symbols $\{L, R\}$ and the map $h : \Sigma_{2+} \to \Lambda$

$$h(\{w_n\}) = \bigcap_{n=0}^{\infty} G^{-n}(\Delta^{w_n}).$$

To each one-sided infinite sequence of symbols (L, R) there corresponds, under h, one and only one point in Λ and vice versa. Actually, it can be proved that h is a homeomorphism (see Katok and Hasselblatt, 1995, p. 81). Moreover, for each $x \in \Lambda$, the sequence corresponding to $G(x)$ can be obtained simply by shifting the sequence corresponding to x one place to the left and dropping the first symbol. Then, denoting again by T_+ the one-sided shift map, the following diagram

$$
\begin{array}{ccc}
\Sigma_+^2 & \xrightarrow{\;T_+\;} & \Sigma_{2+} \\
{\scriptstyle h(\{w_n\})}\Big\downarrow & & \Big\downarrow{\scriptstyle h(\{w_n\})} \\
\Lambda & \xrightarrow[\;G\;]{} & \Lambda
\end{array}
$$

commutes, that is $h \circ T_+ = G \circ h$. We conclude that the map G on Λ is topologically conjugate to the map T_+ on Σ_{2+} and therefore, the dynamics of G on Λ are chaotic.

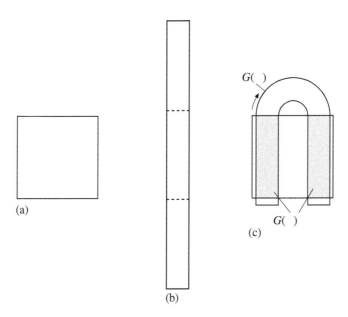

Fig. 6.3 The horseshoe map

6.4 Smale horseshoe

A second example of a map possessing an invariant chaotic set is provided by the celebrated **Smale horseshoe** which is a landmark in the mathematical analysis of chaos. The description below follows Katok and Hasselblatt (1995), pp. 82–3.

Consider a square $\Delta \subset \mathbb{R}^2$ and let $G : \Delta \to \mathbb{R}^2$, be a diffeomorphism whose action on Δ can be described geometrically as follows (it may be helpful to look at figures 6.3 and 6.4 during the explanation). G uniformly contracts Δ in the horizontal direction by a factor $\delta < 1/2$ and uniformly stretches Δ in the vertical direction by a factor $\eta > 2$. The strip thus obtained is then bent and put on top of the square Δ. Part of the original area is mapped outside the square. Assuming that all the deformation consequent to the bending is located in that part, the intersection $G(\Delta) \cap \Delta$ consists of two 'vertical' rectangles Δ^0 and Δ^1 (see figures 6.3(c) and 6.4(a)).

If we now apply the map G to the strips Δ^0 and Δ^1, the points that remain in Δ form four vertical rectangles, Δ^{00}, Δ^{10}, Δ^{11}, Δ^{01}, the width of each rectangle reduced by a factor δ (figures 6.4(b) and (c)).

On the other hand, the intersections $G^{-1}(\Delta) \cap \Delta$ are two 'horizontal' rectangles Δ_0 and Δ_1 (figures 6.4(d) and (e)), and the intersections $G^{-2}(\Delta^i) \cap \Delta$ ($i = 0, 1$) are four horizontal rectangles Δ_{ij} ($i, j = 0, 1$) (figure 6.4(f)). Each

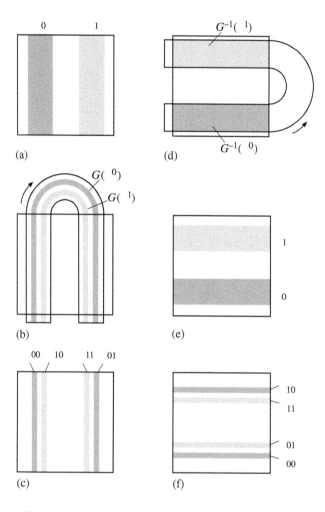

Fig. 6.4 Vertical and horizontal rectangles for the horseshoe map

time the map is applied the height of the rectangles is reduced by a factor $1/\eta$. It can also be verified that $\Delta_i = G^{-1}(\Delta^i)$ $(i = 0, 1)$, $\Delta_{ij} = G^{-2}(\Delta^{ij})$ $(i, j = 0, 1)$, and so on.

We now want to study the properties of the set $\Lambda \subset \Delta$ of points that *never leave* Δ under iterations of G or G^{-1}. The set Λ, assuming that it is nonempty, is by definition invariant (forward and backward) under G and can be formally characterised as

$$\Lambda = \bigcap_{n=-\infty}^{+\infty} G^n(\Delta)$$

where $G^0(\Delta)$ is simply Δ. The set Λ can also be expressed as

$$\Lambda = \Lambda^+ \bigcap \Lambda^-$$

where

$$\Lambda^+ = \bigcap_{n=0}^{+\infty} G^n(\Delta) \qquad \Lambda^- = \bigcap_{n=0}^{+\infty} G^{-n}(\Delta).$$

From our description of the map G we gather that

$$\bigcap_{n=0}^{k} G^n(\Delta)$$

is a set consisting of 2^n thin, disjoint vertical rectangles whose width is equal to δ^n. Analogously

$$\bigcap_{n=0}^{k} G^{-n}(\Delta)$$

is a set consisting of 2^n thin, disjoint horizontal rectangles whose height is equal to η^{-n}.

Taking the horizontal rectangles Δ_0 and Δ_1 as 'building blocks', each of the vertical rectangles $\Delta^{w_1,\dots,w_j}$ has the form

$$\bigcap_{n=1}^{j} G^n(\Delta_{w_n}) \qquad w_n \in \{0,1\}$$

and can be uniquely identified by a j-sequence $(w_1, w_2, \dots, w_j)$ of symbols 0 and 1. In the limit, for $j \to \infty$, as the number of vertical rectangles tends to infinity their width tends to zero (that is, each of them tends to a vertical *segment*) and each of them is identified by a one-sided infinite sequence of 0, 1. The intersection $\bigcap_{n=1}^{\infty} G^n(\Delta)$ is the product of a vertical segment and a Cantor set in the horizontal direction.

Analogously, if we apply the map G backward, each horizontal rectangle can be described by

$$\Delta_{w_0,w_{-1},\dots,w_{-j}} = \bigcap_{n=0}^{j} G^{-n}(\Delta_{w_{-n}}) \qquad w_{-n} \in \{0,1\}$$

and identified uniquely by a one-sided sequence $(w_0, w_{-1}, \dots, w_{-j})$ of 0,1 symbols $(w_{-0} = w_0)$.

In the limit for $j \to \infty$, the number of horizontal rectangles tends to infinity, their height tends to zero, that is, each of them tends to a horizontal segment, and is uniquely identified by a one-sided infinite sequence.

The intersection $\cap_{n=0}^{\infty} G^{-n}(\Delta)$ is the product of a horizontal segment and a Cantor set in the vertical direction. Finally, the intersection of the *vertical* and the *horizontal* Cantor sets, that is, the desired set $\Lambda = \cap_{n=-\infty}^{\infty} G^n(\Delta)$, is itself a Cantor set of points. Each point of the set Λ uniquely corresponds to a bi-infinite sequence of the two symbols. The situation can be summarised by considering the following diagram

$$
\begin{array}{ccc}
\Sigma_2 & \xrightarrow{\ T\ } & \Sigma_2 \\
h(\{w_n\}) \downarrow & & \downarrow h(\{w_n\}) \\
\Lambda & \xrightarrow{\ G\ } & \Lambda
\end{array}
\qquad (6.3)
$$

where Σ_2 is $(\ldots, w_{-1}, w_0, w_1, \ldots)$, the space of bi-infinite sequences of two symbols with $w_n \in \{0,1\}$ for $n \in \mathbb{Z}$, T is the shift map and the map $h(\{w_n\}) = \cap_{n=-\infty}^{\infty} G^n(\Delta_{w_n})$ is a homeomorphism.

If we apply the map G on a point $x \in \Lambda$, we obtain another point $y \in \Lambda$ and the infinite sequence of 0,1 symbols corresponding to y is obtained by shifting the sequence corresponding to x one place to the left. Formally we have

$$
h\left[T(\{w_n\})\right] = G\left[h(\{w_n\})\right]
$$

that is, the diagram (6.3) *commutes*. Therefore the map G on Λ is topologically conjugate to the map T on Σ_2. Since we know the latter to be chaotic, so is the former.

6.5 Tent map and logistic map

The third example of an invariant chaotic set is generated by the (symmetrical) 'tent' map illustrated in figure 6.5.

proposition 6.2 *The tent map* $G_\Lambda : [0,1] \to [0,1]$

$$
G_\Lambda(y) = \begin{cases} 2y, & \text{if } 0 \leq y \leq \frac{1}{2}, \\ 2(1-y), & \text{if } \frac{1}{2} < y \leq 1. \end{cases}
$$

is chaotic on $[0,1]$.

PROOF Consider that the graph of the nth iterate of G_Λ consists of 2^n linear pieces, each with slope $\pm 2^n$. Each of these linear pieces of the graph is defined on a subinterval of $[0,1]$ of length 2^{-n}. Then for any open subinterval J of $[0,1]$, we can find a subinterval K of J of length 2^{-n}, such that the image of K under G_Λ^n covers the entire interval $[0,1]$. Therefore, G_Λ is topologically

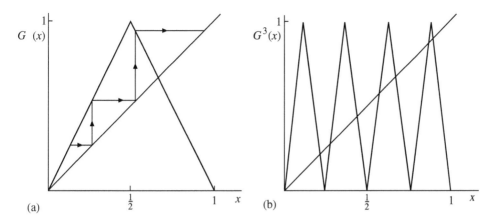

Fig. 6.5 The tent map: (a) G_Λ; (b) G_Λ^3

transitive on $[0, 1]$. This, and the result by Vellekoop and Berglund (1994) discussed in point (5) of remark 6.3 proves the proposition.

remark 6.7 From the geometry of the iterated map G_Λ^n, it appears that the graph of G_Λ^n on J intersects the bisector and therefore G_Λ^n has a fixed point in J. This proves that periodic points are dense in $[0, 1]$. Also for any $x \in J$ there exists a $y \in J$ such that $|G_\Lambda^n(x) - G_\Lambda^n(y)| \geq 1/2$ and therefore G_Λ has SDIC.

This result can be used to show that the logistic map

$$G_4 : [0, 1] \to [0, 1] \qquad G_4(x) = 4x(1 - x)$$

(see figure 6.6) is also chaotic. Consider the map $h(y) = \sin^2(\pi y/2)$. The map h is continuous and, restricted to $[0, 1]$, it is also one-to-one and onto. Its inverse is continuous and h is thus a homeomorphism. Consider now the diagram

$$
\begin{array}{ccc}
[0, 1] & \xrightarrow{\;G_\Lambda\;} & [0, 1] \\
{\scriptstyle h(y)}\downarrow & & \downarrow{\scriptstyle h(y)} \\
[0, 1] & \xrightarrow[\;G_4\;]{} & [0, 1]
\end{array}
$$

where G_Λ is the tent map. Recalling the trigonometric relations:

(i) $\sin^2(\theta) + \cos^2(\theta) = 1$;
(ii) $4\sin^2(\theta)\cos^2(\theta) = \sin^2(2\theta)$;
(iii) $\sin(\pi - \theta) = \sin(\theta)$;

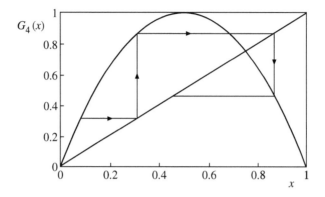

Fig. 6.6 The logistic map G_4

we can see that the diagram commutes. Hence, the map G_4 is topologically conjugate to G_Λ and therefore, its dynamics on $[0, 1]$ are chaotic. We shall use again the equivalence relation between the logistic and the tent maps when we discuss their ergodic properties in chapters 9 and 10.

6.6 Doubling maps

Next we discuss the doubling (or double angle) map, illustrated in figure 6.7 on the unit circle (also known as the 'squaring map').[2] In the multiplicative notation of the unit circle we have

$$G_C : S^1 \to S^1$$
$$S^1 = \{z \in \mathbb{C} | |z| = 1\} = \{e^{i2\pi x} | x \in \mathbb{R}\}$$
$$G_C(z) = z^2.$$

proposition 6.3 *The doubling map on the circle is chaotic on S^1.*

PROOF To prove transitivity of G_C, consider any two open subsets of S^1 (open arcs) U and V. If l is the length of U then G_C doubles l at each iteration and therefore, for some positive integer k, $G_C^k(U)$ will cover S^1 and consequently $G_C^k(U) \cap V \neq \emptyset$. On the other hand, k can be chosen so that G_C^k maps arbitrarily small arcs to S^1. Therefore, G_C has sensitive dependence on initial conditions and consequently is chaotic. (The proof that the set of periodic points is dense in S^1 is a bit more complicated

---

[2]It is 'doubling' in the additive notation of the unit circle. It is 'squaring' in the multiplicative notation.

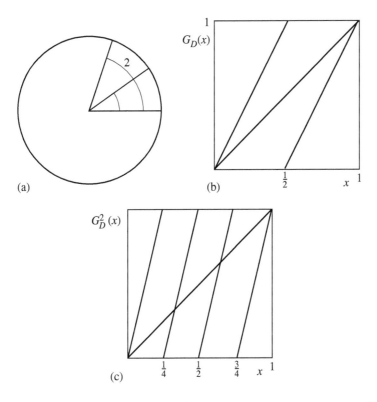

Fig. 6.7 Doubling map: (a) on S^1; (b) on the unit interval; (c) G_D^2

and we refer the reader to the literature, for example, Elaydi, 2000, p. 108, example 3.5.)

Consider now the (discontinuous) doubling map *on the interval* (with the ordinary Euclidean metric)

$$G_D : [0, 1] \rightarrow [0, 1] \qquad G_D(x) = \begin{cases} 2x, & \text{if } 0 \le x \le \frac{1}{2}, \\ 2x - 1, & \text{if } \frac{1}{2} < x \le 1. \end{cases}$$

proposition 6.4 *The map G_D is chaotic on the interval $[0, 1]$.*

PROOF To prove this proposition, consider first that the graph of G_D^n consists of 2^n linear pieces, each with slope 2^n, and defined on an interval of length 2^{-n}. Then, the same argument used in proposition 6.2 for the tent map can be used to prove that G_D is topologically transitive on $[0, 1]$. Density of periodic orbits and SDIC can be established by using the same argument used in remark 6.7 for the tent map.

remark 6.8 Notice that the doubling map *on the circle* is expanding, whereas the tent map and the doubling map *on the interval* are piecewise expanding, according to the following definition.

definition 6.4 *A continuous map G on a metric space M is called* **expanding** *if for some $\rho > 1$ and $\epsilon > 0$, and for every $x, y \in M$ with $x \neq y$ and $d(x, y) < \epsilon$, we have*

$$d[G(x), G(y)] > \rho d(x, y)$$

where d is a distance function on M. A **piecewise expanding** *map of the interval is a piecewise continuous map on $[0, 1]$ (or $[0, 1)$) such that, on each piece we have $|G'(x)| > 1$.*

Piecewise expanding maps of the interval are characterised by a strong form of chaotic behaviour called 'Bernoullian', which we discuss in chapters 9 and 10. For a detailed discussion of the interesting properties of these maps, see Boyarsky and Góra (1997).

6.7 Chaotic attractors

In the previous sections we discussed examples of simple maps which have chaotic dynamics in the sense that there exist invariant sets on which these maps are topologically transitive and have SDIC. We now turn to the question of attractiveness of chaotic sets.

Attractiveness is an empirically relevant property because it is closely related to observability in real experiments or in computer-generated numerical simulations. In either case, we expect the *observed* motion of the system to be subjected to perturbations, owing to experimental noise or to approximations in computer operations. Thus, the observability of chaos requires that the set of initial conditions whose orbits converge to the chaotic set should not be *exceptional* in a sense to be made precise. A stronger requirement is that the chaotic set should be stable under small random perturbations.

We now specifically pose the following question. Are the chaotic invariant sets studied in the previous sections attractors? Because there are different characterisations of attractors, we expect different answers according to which definition we choose. We take the aforementioned models in turn.

SMALE HORSESHOE The invariant set Λ is a set of points of Lebesgue measure zero. All orbits starting from points in the set $\Delta \setminus \Lambda$ (that is, almost all points in the square Δ) leave the square after a sufficiently large number of iterates of the map G. Thus, Λ is not an attractor in the sense of either definition 4.10 or definition 4.11.

TENT (G_Λ) AND LOGISTIC (G_4) MAPS For these maps, the interval $[0,1]$, as a subset of $\mathbb{R}$, is an attractor in the weak sense of definition 4.11 and its stable set is the attractor itself. However, to discuss stability of the set $[0,1]$, we must consider orbits that start from initial points outside $[0,1]$. For example, if we consider the map G_4 from $\mathbb{R}$ to $\mathbb{R}$, we can see immediately that $[0,1]$ repels every point outside itself and consequently it is not an attractor in the sense of 4.10 (cf. Milnor, 1985,[3] p. 184, example 3). Analogous considerations could be made for the tent map.

LOGISTIC MAP WITH $\mu > 2 + \sqrt{5}$ For this map and its invariant chaotic set Λ, we can repeat the same considerations we have just made for the Smale horseshoe. Thus, Λ is not an attractor according to either definition 4.10 or 4.11.

EXTENDED DOUBLING MAP A simple case of an asymptotically stable chaotic invariant set is provided by an extension of the doubling map of the circle $G_C(z) = z^2$, $z \in S^1$ discussed in section 6.6. Consider the map

$$G_\theta : [S^1 \times \mathbb{R}] \to [S^1 \times \mathbb{R}] \qquad G_\theta(z,x) = (z^2, cx) \qquad (6.4)$$

where $z = e^{i2\pi\theta}$ and $\theta, x \in \mathbb{R}$, $0 < c < 1/2$. The state space $S^1 \times \mathbb{R}$ of (6.4) is a cylinder as shown in figure 6.8(a).

Notice that under iteration of G_θ, starting on any point on the cylinder the x coordinate will converge to zero as $n \to +\infty$ for all c in the prescribed interval. Therefore all orbits tend asymptotically to the invariant circle $S^1 \times \{0\}$. The dynamics on $S^1 \times \{0\}$ is determined by the map G_C, and we know that it is chaotic. Thus $S^1 \times \{0\}$ is a chaotic attractor for G_θ.

System (6.4) can also be represented as a map of the plane, by the following coordinate change

$$\theta = \frac{\ln z}{i2\pi}$$
$$r = e^x$$

whence

$$\theta_{n+1} = \frac{\ln z_{n+1}}{i2\pi} = \frac{\ln z_n^2}{i2\pi} = 2\theta_n \bmod 1 \qquad (6.5)$$

[3]Notice that Milnor refers to the map $f(x) = 2x^2 - 1$ and the corresponding invariant interval $[-1,1]$. But $f(x)$ on $[-1,1]$ is dynamically equivalent to G_4 on $[0,1]$.

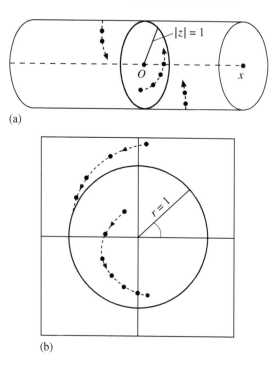

(a)

(b)

Fig. 6.8 The extended doubling map: (a) on the cylinder; (b) on the plane

$$r_{n+1} = e^{x_{n+1}} = e^{cx_n} = r_n^c. \tag{6.6}$$

Thus r and θ are the polar coordinates of points in $\mathbb{R}^2$. From (6.6) we gather that the unit circle is attracting and the dynamics on it is governed by the doubling map (6.5). The dynamics on $S^1 \times \mathbb{R}$ are illustrated in figure 6.8(a), and those on the plane in (b).

6.8 The Lorenz model

Although there is plenty of numerical evidence of chaotic behaviour in dynamical models arising from a variety of problems in different fields of applications, apart from the cases of one-dimensional, noninvertible maps, there are few rigorous proofs that specific mathematical models possess chaotic attractors as characterised in one or the other of the above definitions, and those proofs are mostly restricted to artificial examples unlikely to arise in typical applications. We now turn to an example of a chaotic attractor derived from a celebrated model first discussed in 1963 by Edmund Lorenz to provide a mathematical description of atmospheric turbulence. Lorenz's investigation was enormously influential and stimulated a vast literature in the years that followed. Its extraordinary success was in part owing to the fact that it showed how computer technology, which at that moment had only recently been developed, could be effectively used to study nonlinear dynamical systems. In so doing, Lorenz's work provided strong numerical evidence that a low-dimensional system of differential equations with simple nonlinearities could generate extremely complicated orbits.

And yet, despite over three decades of sustained mathematical analysis of the model, we are still lacking a conclusive, rigorous proof that what we see on the computer screen when we simulate the orbits of the Lorenz equations is a truly chaotic attractor. We do have results for a *geometrical model* which is derived from the original one by means of some convenient assumptions and is compatible with the observed numerical results. We recall the main analytical and numerical results available for the original Lorenz model before sketching the associated geometric model.

The original Lorenz model is defined by the following three differential equations:

$$\dot{x} = -\sigma x + \sigma y$$
$$\dot{y} = -xz + rx - y \tag{6.7}$$
$$\dot{z} = xy - bz$$

where $x, y, z \in \mathbb{R}$; $\sigma, r, b > 0$. System (6.7) is symmetrical under the transformation $(x, y, z) \rightarrow (-x, -y, z)$. Its equilibria are the (real) solutions to the equation

$$\bar{x}\left[b(r-1) - \bar{x}^2\right] = 0$$

obtained by setting $\dot{x} = \dot{y} = \dot{z} = 0$. For $0 < r < 1$ the only equilibrium is $E_1 : (0, 0, 0)$. For $r > 1$ there exist two other equilibria, namely $E_2 : (+\sqrt{b(r-1)}, +\sqrt{b(r-1)}, r-1)$ and $E_3 : (-\sqrt{b(r-1)}, -\sqrt{b(r-1)}, r-1)$.

Local stability of the equilibria can be ascertained by evaluating the sign of the real part of the eigenvalues of the Jacobian matrix

$$Df(x) = \begin{pmatrix} -\sigma & \sigma & 0 \\ r-z & -1 & -x \\ y & x & -b \end{pmatrix}$$

at equilibrium. The trivial equilibrium E_1 is stable for $0 < r < 1$. Bifurcations of the other two equilibria can be located by finding the combinations of the parameter values for which they become nonhyperbolic, that is, for which the auxiliary equation

$$\lambda^3 + (b + \sigma + 1)\lambda^2 + b(r + \sigma)\lambda + 2b\sigma(r-1) = 0 \tag{6.8}$$

(calculated in either fixed point) has a solution $\lambda = 0$ or an imaginary pair $\lambda_{1,2} = \pm i\beta$ $\beta \in \mathbb{R}$. Substituting $\lambda = 0$ into (6.8) gives $2b\sigma(r-1) = 0$ whence $r = 1$. At $r = 1$ there is a pitchfork bifurcation, the old stable equilibrium becomes unstable and two new stable equilibria are born. Notice that for $r > 1$ the Jacobian matrix at the origin has two negative eigenvalues and one positive eigenvalue.

Substituting $\lambda = i\beta$ into (6.8) and equating both real and imaginary parts to zero we have

$$-(b + \sigma + 1)\beta^2 + 2b\sigma(r - 1) = 0$$
$$-\beta^3 + b(r + \sigma)\beta = 0$$

whence

$$\beta^2 = \frac{2b\sigma(r - 1)}{b + \sigma + 1} = b(r + \sigma).$$

Since β is real, $\beta^2 > 0$ implying $r > 1$. Then, as long as $\sigma > b + 1$, there is a Hopf bifurcation at $r_H = \frac{\sigma(\sigma+b+3)}{\sigma-b-1}$.

In numerical analyses of the Lorenz model, the typical parameter configuration is $\sigma = 10$ $b = 8/3 = 2.\bar{6}$, giving $r_H \approx 24.74$. Because of the symmetry of the nontrivial equilibria E_2 and E_3, the stability properties are equivalent and the same bifurcations take place for the same parameter values for both equilibria. The situation is illustrated in figure 6.9. At $r_H \approx 24.74$ there is a pair of symmetrical, *subcritical Hopf bifurcations* (see chapter 5, section 5.3). This means that when r is *decreased* through r_H, two symmetric *unstable* periodic orbits are created, and they are destroyed at some value $1 < r < r_H$. For values of $r \in (1, r_H)$, E_2 and E_3 are stable, but there is numerical evidence that for $r > r_c \approx 24.06$ a chaotic attractor coexists with the stable equilibria. Beyond r_H these equilibria become unstable.

For values of $r > r_H$ the celebrated *Lorenz attractor* (the so-called 'butterfly', see figure 6.10) is observed numerically. A typical orbit coils around one of the unstable fixed points E_2, E_3, eventually crosses over and coils around the other, passing from one 'wing' of the butterfly to the other again and again. There is also strong evidence of sensitive dependence on initial conditions: two orbits starting from two initially close points after a finite time will be distant, even on opposite wings of the butterfly.

As mentioned before, a full mathematical analysis of the observed attractor is still lacking. However, some of the attractor's properties have been established through a combination of numerical evidence and theoretical

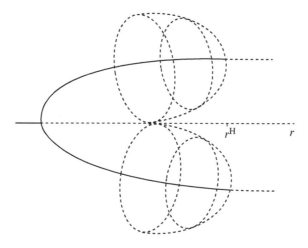

Fig. 6.9 Lorenz model with two subcritical Hopf bifurcations

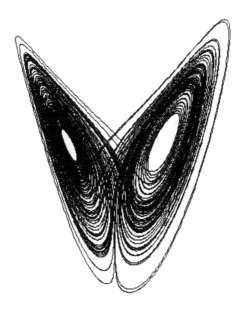

Fig. 6.10 The Lorenz attractor

arguments which we present below.[4] First, however, consider the following three facts about system (6.7):

---

[4]For a discussion of the Lorenz attractor and related questions by one of the most prominent authors in the field, see Shilnikov (1997).

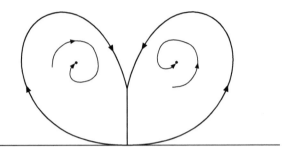

Fig. 6.11 Homoclinic orbits in the Lorenz model

(i) The trace of the Jacobian matrix,

$$\text{tr}[Df(x, y, z)] = \frac{\partial \dot{x}}{\partial x} + \frac{\partial \dot{y}}{\partial y} + \frac{\partial \dot{z}}{\partial z} = -(b + \sigma + 1) < 0$$

is constant and negative along orbits. Thus, any three-dimensional volume of initial conditions is contracted along orbits at a rate equal to $\gamma = -(b + \sigma + 1) < 0$, that is to say, the system is dissipative (cf. the appendix to chapter 4, 128).

(ii) It is possible to define a trapping region (see definition 4.6) such that all orbits outside of it tend to it, and no orbits ever leave it. To see this, consider the function[5]

$$V(x, y, z) = x^2 + y^2 + (z - r - \sigma)^2 = K^2(r + \sigma)^2 \qquad (6.9)$$

defining a sphere with centre in $(x = y = 0; z = \sigma + r)$ and radius $[K(\sigma + r)]$. Taking the total time derivative of (6.9) and substituting from system (6.7), we have

$$\dot{V}(x, y, z) = -2\sigma x^2 - 2y^2 - 2b\left(z - \frac{r + \sigma}{2}\right) + b\frac{(r + \sigma)^2}{2}. \qquad (6.10)$$

$\dot{V} = 0$ defines an ellipsoid outside of which $\dot{V} < 0$. For a sufficiently large value of the radius (for sufficiently large K, given r and σ), the sphere (6.9) will contain all three fixed points and all orbits on the boundary of the sphere will point inwards. Consequently, system (6.7) is 'trapped' inside the sphere.

(iii) Numerical evidence indicates that for $r \in (13.8, 14)$, there exist two symmetric **homoclinic orbits** (that is, orbits that connect a fixed point to itself) asymptotically approaching the origin for $t \to \pm\infty$, tangentially to the z-axis, see the sketch in figure 6.11.

[5]See Neimark and Landa (1992), pp. 201–2. A slightly different function defining a trapping region for the Lorenz model is used, for example, by Glendinning (1994), pp. 49–50.

Keeping in mind these three facts, the situation can be investigated by constructing a geometric model which, under certain hypotheses, provides a reasonable approximation of the dynamics of the original model with 'canonical' parameter values $\sigma = 10$, $b = 8/3$, and $r > r_H$. Since the construction of the model involves many technical details, we recall here only the main features, referring the reader to the specialised literature for a complete discussion (see Sparrow (1982), Guckenheimer and Holmes (1983)). The following presentation is based on Glendinning (1994), pp. 362–7 and Robinson (1999), pp. 347–53.

We first consider a system of differential equations in $\mathbb{R}^3$ depending on a parameter μ with the following properties:

(i) for a certain value μ_h of the parameter, there exists a pair of symmetrical homoclinic orbits, asymptotically converging to the origin, tangential to the positive z-axis;

(ii) the origin is a saddle-point equilibrium and the dynamics in a neighbourhood N of the equilibrium, for μ in a neighbourhood of μ_h, is approximated by the system

$$\dot{x} = \lambda_1 x$$
$$\dot{y} = \lambda_2 y \qquad\qquad (6.11)$$
$$\dot{z} = \lambda_3 z$$

where $\lambda_2 < \lambda_3 < 0 < \lambda_1$ and $-\lambda_3/\lambda_1 < 1$;

(iii) the system is invariant under the change of coordinates $(x, y, z) \rightarrow (-x, -y, z)$.

Under these conditions, for $(x, y, z) \in N$ and μ near μ_h, it is possible to construct a two-dimensional cross-section Σ, such that the transversal intersections of orbits with Σ define a two-dimensional Poincaré map $P : \Sigma \rightarrow \Sigma$. For values of $|x|$ and $|\mu - \mu_h|$ sufficiently small, the dynamics of P can be further approximated by a one-dimensional, noninvertible map $G_\mu[-a, a] \setminus \{0\} \rightarrow \mathbb{R}$ defined on an interval of the x-axis, but not at $x = 0$.

A typical formulation of the map G_μ is

$$G_\mu(x) = \begin{cases} a\mu + cx^\delta, & \text{if } x > 0; \\ -a\mu - c|x|^\delta, & \text{if } x < 0; \end{cases}$$

where $a < 0$, $c > 0$, $\delta = -\lambda_3/\lambda_1$, $1 > \delta > 0$.

Assuming that the one-dimensional approximation remains valid for values of the parameter μ outside the neighbourhood of $\mu = \mu_h$ (that is, outside the neighbourhood of the homoclinic orbit), values of $\mu > \mu_h$ can be

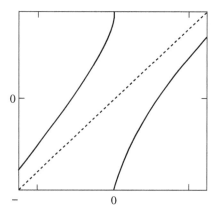

Fig. 6.12 One-dimensional map for the Lorenz model

chosen so that there exists a closed interval $[-\alpha, \alpha]$ with $\alpha > 0$ such that $G_\mu[-\alpha, \alpha] \setminus \{0\} \to [-\alpha, \alpha]$ and

$$\lim_{x \to 0^-} G_\mu(x) = \alpha > 0 \qquad \lim_{x \to 0^+} G_\mu(x) = -\alpha < 0.$$

Then $G'_\mu(x) > 1$ for $x \in [-\alpha, \alpha]$ $x \neq 0$, and $\lim_{x \to 0^\pm} G'_\mu(x) = +\infty$. The map G_μ, which looks like a distorted version of the doubling map on the interval is depicted in figure 6.12.

Because $G'_\mu(x) > 1$ for all $x \in [-\alpha, \alpha]$, G_μ is a piecewise expanding map and has therefore sensitive dependence on initial conditions. There are no fixed points or stable periodic points and most orbits on the interval $[-\alpha, 0) \cup (0, \alpha]$ are attracted to a chaotic invariant set. Although increasing μ beyond the *homoclinic value* μ_h leads to stable chaotic motion, if we take μ very large, the system reverts to simpler dynamical behaviour and stable periodic orbits reappear. This phenomenon is by no means exceptional in nonlinear systems. (For an example in the mathematical economics literature, see Invernizzi and Medio, 1991.)

remark 6.9 The idea that some essential aspects of the dynamics of the original system (6.7) could be described by a one-dimensional map was first put forward by Lorenz himself. In order to ascertain whether the numerically observed attractor could be periodic rather than chaotic, he plotted successive maxima of the variable z along an orbit on the numerically observed attractor. In doing so, he discovered that the plot of z_{n+1} against z_n has a simple shape, as illustrated in figure 6.13. The points of the plot lie almost exactly on a curve whose form changes as the parameter r varies.

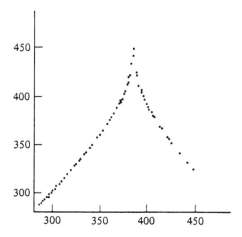

Fig. 6.13 Successive maxima for the Lorenz attractor

Setting $\sigma = 10$, $b = 8/3$, $r = 28$ (the traditional 'chaotic values'), we obtain a curve resembling a distorted tent map. It has slope everywhere greater than one in absolute value so, again, it approximates a piecewise expanding map. For such a map there cannot be stable fixed point or stable periodic orbits and for randomly chosen initial values, orbits converge to a chaotic attractor.

Exercises

6.1   From exercise 5.8(a) we know that for the map $G(x_n) = \mu - x_n^2$ there is a single stable branch of fixed points over $\frac{-1}{4} < \mu < \frac{3}{4}$ which undergoes a flip bifurcation at $\mu = \frac{3}{4}$. Thereafter follows a period-doubling route to chaos (see a generic representation in figure 5.12). Consider two cases: $\mu = 1$, for which a period-2 cycle $\{0, 1\}$ exists; $\mu = 1.9$, for which the dynamics of G numerically appears to be chaotic:

    (a) Calculate four iterations for each of two orbits starting at initial values $\hat{x}_0 = 0.2$ and $x_0 = 0.1$ (rounding off to 4 decimal places) and comment on how the distance between orbits changes.

    (b) Notice that nearby orbits diverge but then come close again in the case of presumed chaotic orbits. To see this, iterate once again for the case $\mu = 1.9$.

6.2   Prove that the infinite sum in (6.2) converges and is ≤ 3.

6.3   Prove that T, as defined in section 6.2, is invertible but T_+ is not.

6.4   Consider the following problems involving bi-infinite sequences:

   (a) Let $s = $ {bi-infinite sequence of $\{0\}$} and $\bar{s} = $ {bi-infinite se-
       quence of $\{1\}$}. Evaluate the distance between s and $\bar{s}$.

   (b) Let $s = $ {bi-infinite sequence of $\{0\ 1\}$} and $\bar{s} = $ {bi-infinite
       sequence of $\{1\ 0\}$}, $s_0 = 0$, $\bar{s}_0 = 1$. Evaluate the distance
       between s and $\bar{s}$.

   (c) Given the following bi-infinite sequences:

$$s = \{ \text{ all zeros } s_{-1}s_0s_1 \text{ all zeros } \}$$
$$\bar{s} = \{ \text{ all zeros } 0\ 1\ 0 \text{ all zeros } \}$$
$$\hat{s} = \{ \text{ all zeros } 0\ 1\ 1 \text{ all zeros } \}$$
$$\tilde{s} = \{ \text{ all zeros } 0\ 0\ 1 \text{ all zeros } \}$$

   calculate the distances $(\bar{s}, \hat{s})$, $(\hat{s}, \tilde{s})$, $(\bar{s}, \tilde{s})$.

6.5   Write the one-sided sequence s^* described in proposition 6.1(ii),
       through to the block $\{0001\}$. Let s be the one-sided sequence of
       all zeros:

   (a) Find n and k such that $d[T_+^n(s^*), s] \le \frac{1}{32}$, $d[T_+^n(s^*), s] \le \frac{1}{256}$.

   (b) Let $t = T_+^2(s^*)$ and $\delta = 1$ in definition 6.1. Give the first 10
       values of n for which $d[T_+^n(t), T_+^n(s)] \ge 1$.

6.6   Prove that the set of periodic points of G_Λ is dense on $[0, 1]$ and that
       G_Λ has sensitive dependence on initial conditions.

6.7   Show that the map

$$F : [-1, 1] \to [-1, 1] \qquad F(x) = 2x^2 - 1$$

   is topologically conjugate to the tent map.

6.8   Show that the maps

$$F : [-1, 1] \to [-1, 1] \qquad F(x_n) = 2x_n^2 - 1$$
$$G : [0, 1] \to [0, 1] \qquad G(y_n) = 4y_n(1 - y_n)$$

   are topologically conjugate.

6.9   Show that the following pairs of maps are topologically conjugate
       by the given homeomorphism:

   (a) the logistic map at $\mu = 4$, G_4, and the tent map, G_Λ, by the
       homeomorphism as given in section 6.5, $h(x) = \sin^2\left(\frac{\pi x}{2}\right)$;

   (b) (Alligood et al., 1997, p. 116) the same maps as for (a), and
       $h(x) = \frac{1 - \cos(\pi x)}{2}$ (recall that $\sin^2(\theta) = \frac{1}{2}[1 - \cos(2\theta)]$);

(c) (Elaydi, 2000, pp. 129–30, example 3.10) the logistic map and the quadratic map $F(x_n) = ax_n^2 + bx_n + c$, $a \neq 0$, $h(x) = \frac{-\mu x}{a} + \frac{\mu - b}{2a}$.

6.10   Consider the differential equation in $\mathbb{R}$

$$\dot{x} = f(x) = 4x(1 - x) \qquad\qquad (i)$$

and its (Euler) discrete-time approximation

$$x_{t+T} = x_t + Tf(x_t) \qquad t \in \mathbb{R}, T > 0. \qquad (ii)$$

Find the exact solution of (i) as a function of t. As $T \to 0$, equation (ii) tends to (i) (why?) and therefore we expect that, for small T, the dynamics of (ii) should be a good approximation of that of (i). Try now to construct numerically a bifurcation diagram of (ii) for increasing values of T and discuss the results. What is the lowest value of T for which (ii) has (a) period-2 cycle, (b) period-4 cycle?

6.11   Consider the following model, known in economics as the Solow growth model

$$\dot{k} = sf(k) - (n + \delta)k \qquad\qquad (i)$$

where $k \in \mathbb{R}_+$ denotes the capital stock, $f : \mathbb{R}_+ \to \mathbb{R}_+$ is the production function assumed to be concave, and the parameters are: $0 \leq s \leq 1$, the propensity to save; $\delta > 0$, the rate of capital depreciation; $n > 0$, the rate of population growth (n and δ are typically of the order 10^{-2}, 10^{-1}, respectively). Identify the equilibrium points and discuss their stability.

Rewrite system (i) in discrete-time form

$$k_{n+1} = sf(k_n) - (n + \delta)k_n. \qquad\qquad (ii)$$

Could system (i) be characterised by complex behaviour? Could system (ii)? Explain your answers.

6.12   (Jackson, 1990, p. 181, exercise 4.7) For the special case $\mu = 4$, the logistic map has a closed-form solution

$$x(n) = \sin^2(2^n \pi \theta) \qquad\qquad (i)$$

where θ depends on the initial condition.

(a) Show that (i) is a solution for

$$x_{n+1} = 4x_n(1 - x_n).$$

(b) Show that the set of initial conditions corresponding to nonperiodic solutions has Lebesgue measure one on the set $[0, 1]$.

(c) Show that the set corresponding to periodic solutions is dense in $[0, 1]$. (*Hint*: consider that periodicity or aperiodicity of solutions depends on whether the initial condition is a rational or irrational number.)

7

Characteristic exponents, fractals, homoclinic orbits

7.1 Lyapunov characteristic exponents

Having thus discussed some 'canonical' types of chaotic sets and attractors, it is now time to provide a more rigorous characterisation of Lyapunov characteristic exponents as a measure of exponential divergence of orbits. Let us first discuss LCEs for maps. We begin with a simple, one-dimensional setting and then generalise it. Consider the map

$$x_{n+1} = G(x_n)$$

with $G : U \to \mathbb{R}$, U being an open subset of $\mathbb{R}$, G is continuously differentiable. The LCE, a measure of the divergence of nearby orbits, is calculated by following the evolution in time of two orbits originating from two nearby points x_0 and $\hat{x}_0$. At the nth iteration they are separated by an amount

$$|\hat{x}_n - x_n| = |G^n(\hat{x}_0) - G^n(x_0)| = |T(\hat{x}_0, x_0)|.$$

Taking x_0 as a constant, expanding $T(\hat{x}_0)$ in a Taylor series around $\hat{x}_0 = x_0$, and retaining only the first-order term, we have

$$\hat{x}_n - x_n \approx \left. \frac{dG^n}{dx} \right|_{\hat{x}_0 = x_0} (\hat{x}_0 - x_0) = [G'(x_0)G'(x_1) \cdots G'(x_{n-1})] (\hat{x}_0 - x_0)$$

where, $x_i = G^i(x_0)$ and, providing the derivative is different from zero, the approximation can be made arbitrarily accurate by taking $|(\hat{x}_0 - x_0)|$ sufficiently small. Asymptotically, we have

$$\lim_{n \to \infty} |\hat{x}_n - x_n| \approx e^{n\lambda(x_0)} |\hat{x}_0 - x_0|$$

where the quantity

$$\lambda(x_0) = \lim_{n \to \infty} \frac{1}{n} \ln |G'(x_0)G'(x_1) \cdots G'(x_{n-1})|$$

$$= \lim_{n \to \infty} \frac{1}{n} \sum_{j=0}^{n-1} \ln |G'(x_j)| \qquad (7.1)$$

(assuming the limit exists) is called the **Lyapunov characteristic exponent** (LCE). The interpretation of $\lambda(x_0)$ is straightforward: it is the (local) average, asymptotic, exponential rate of divergence of nearby orbits.[1]

For one-dimensional maps we have, of course, a single LCE. If the orbit of x_0 converges to a stable periodic orbit (recall that a fixed point is periodic of period 1) then $\lambda(x_0) < 0$. If the orbit of x_0 is an unstable periodic orbit, then $\lambda(x_0) > 0$. If the orbit of x_0 is not periodic or asymptotically periodic (but bounded) and $\lambda(x_0) > 0$, then the orbit is chaotic. If the orbit of x_0 converges to a quasiperiodic (aperiodic, nonchaotic) orbit, $\lambda(x_0) = 0$. Finally, if the orbit of x_0 converges to a periodic orbit that is nonhyperbolic (at bifurcation points) $\lambda(x_0) = 0$. Taking the logistic map as an example, if $\mu = 3.2$ a stable period-2 cycle attracts most orbits and, for a generic x_0, we have $\lambda(x_0) < 0$; if $\mu = 1 + \sqrt{6}$, at which value the period-2 cycle loses stability, for a generic x_0 $\lambda(x_0) = 0$; if $\mu = 4$, for a generic x_0 $\lambda(x_0) = \ln 2 > 0$ and the typical orbit is chaotic.

remark 7.1 For a one-dimensional map G, when the orbit of x_0 is attracted to a periodic orbit of period k, $k \geq 1$, the corresponding LCE is equal to the logarithm of the absolute value of $\partial G^k(x^*)/\partial x_n$ which can be thought of as the eigenvalue of the matrix $DG^k(x^*)$ (x^* any periodic point). In the stable case, $|\partial G^k(x^*)/\partial x_n| < 1$ and consequently the LCE is negative.

In the multi-dimensional case, the calculation of LCEs can be done in a similar, albeit more complex manner. We discuss the discrete-time case and add a few comments on flows later. Consider again the map

$$x_{n+1} = G(x_n)$$

where we now have $x_n = (x_n^1, \ldots, x_n^m) \in \mathbb{R}^m$ and G is a vector of smooth functions $(G_1, G_2, \ldots, G_m)$ such that $G_l : U \subset \mathbb{R}^m \to \mathbb{R}$, $(l = 1, \ldots, m)$. Consider the orbits originating at a point x_0 in the state space and a nearby point $\hat{x}_0$. The rate of change of the distance between them (where distance

---

[1] It is local since we evaluate the rate of separation in the limit for $\hat{x}_0 \to x_0$. It is asymptotic since we evaluate it in the limit of an indefinitely large number of iterations — assuming that the limit exists. The presence of two limits creates some problems in making this argument perfectly rigorous. Thus, one sometimes speaks of asymptotic behaviour of *infinitesimal* initial distances. (See, on this point, Robinson, 1999, p. 89.)

is measured by $\| \cdot \|$, a norm on $\mathbb{R}^m$ such as the Euclidean norm), evolves under the action of the map G according to the ratio

$$\|G^n(\hat{x}_0) - G^n(x_0)\|/\|\hat{x}_0 - x_0\|.$$

As we want to study the time evolution of nearby orbits, we take the limit for $\hat{x}_0 \to x_0$, giving:

$$\frac{\|DG^n(x_0)w\|}{\|w\|} = \frac{\|\prod_{i=0}^{n-1} DG(x_i)w\|}{\|w\|} \qquad (7.2)$$

where $x_i = G^i(x_0)$; for each $(i = 0, 1, \ldots, n-1)$ $DG(x_i)$ is a $(m \times m)$ matrix with typical element $[\partial G_l / \partial x_i^j]$ and x_i^j $(j = 1, \ldots, m)$ is the jth element of the vector x_i;

$$DG^n(x_0) = DG(x_0)DG(x_1) \cdots DG(x_{n-1});$$

and w is a vector in the *tangent space* at x_0.

We can now define the **LCEs of the vector w** (which in general will also depend on x_0) as follows:

$$\lambda(x_0, w) = \lim_{n \to \infty} \frac{1}{n} \ln \frac{\|DG^n(x_0)w\|}{\|w\|} \qquad (7.3)$$

assuming that this limit exists. The intuitive interpretation of LCEs given above for a one-dimensional map, carries over to the multi-dimensional case.

First of all, consider that, because the Lyapunov exponents do not depend on the length of the vector w, we can always simplify (7.3) by setting $\|w\| = 1$. Next, consider that under the action of the matrix $DG^n(x_0)$ the vector w will be stretched, or contracted (and rotated). Now adopting the vector norm $\| \cdot \| = (\cdot, \cdot)^{\frac{1}{2}}$, where $(\cdot, \cdot)$ denotes the scalar product operation, the norm (the length) of the image of w under the matrix $DG^n(x_0)$ will be

$$\|DG^n(x_0)w\| = \left(DG^n(x_0)w, DG^n(x_0)w \right)^{\frac{1}{2}}$$

$$= \left(w^T [DG^n(x_0)^T DG^n(x_0)]w \right)^{\frac{1}{2}} \qquad (7.4)$$

where we can take square roots because the matrix $[DG^n(x_0)^T DG^n(x_0)]$ is positive definite. Thus, the matrix $[DG^n(x_0)^T DG^n(x_0)]^{1/2}$ determines to what extent lengths are expanded (contracted) under the action of $DG^n(x_0)$. The average amount of stretching (contraction) per iteration is given by $(DG^n(x_0)w, DG^n(x_0)w)^{1/2n}$. In the limit for $n \to \infty$ (if it exists) the eigenvalues of the matrix

$$\Lambda(x_0) \equiv \lim_{n \to \infty} \left[DG^n(x_0)^T DG^n(x_0) \right]^{\frac{1}{2n}} \qquad (7.5)$$

are called **Lyapunov characteristic numbers** and their logarithms, the **Lyapunov characteristic exponents** defined by (7.3), measure the asymptotic, average, exponential rate of stretching (contraction) of the vector w.

remark 7.2 From formula (7.5), we notice again that LCEs depend not only on the vector w in the tangent space, but also on the initial condition x_0. We shall see in chapter 9 under what conditions LCEs are the same for 'most' initial points. This will also give a more precise meaning to the phrase 'typical orbit'.

We have so far discussed LCEs for maps. The discussion can be extended to continuous time systems of differential equations $\dot{x} = f(x)$ $x \in \mathbb{R}^m$ by considering its flow map $\phi(t, x)$. The Lyapunov numbers and exponents can be defined in the same way as for maps, replacing the matrix $DG^n(x)$ of equation (7.3) with the matrix of partial derivatives of $\phi(t, x)$ with respect to x, $D_x\phi(t, x)$, evaluated along the orbits of the system starting at x. Then, the LCEs are defined as

$$\lambda(x, w) = \lim_{t \to \infty} \frac{1}{t} \ln \|D_x\phi(t, x)w\| \qquad t \in \mathbb{R}.$$

remark 7.3 Notice that the time-dependent matrix $B(t) \equiv D_x\phi(t, x)$ satisfies the matrix differential equations

$$\frac{d}{dt}B(t) = A(t)B(t) \qquad A(t) \equiv D_x f[\phi(t, x)]. \tag{7.6}$$

Equation (7.6) is known as a **variational equation** which, intuitively speaking, indicates that the matrix $A(t)$ represents the action of the vector field f for small displacements along the vector w.

Except for a few special cases, we cannot evaluate LCEs of nonlinear maps or flows exactly. In most cases, therefore, we have to resort to numerical computation. There exist several, more or less accurate and efficient numerical methods to compute LCEs. We will not discuss them here[2] in any detail but only provide a few general considerations. Because most vectors will have some components in the direction of the eigenvector corresponding to the *largest* eigenvalue, if we choose w at random and numerically calculate the quantity

$$\lambda(x_0) = \lim_{n \to \infty} \frac{1}{n} \ln \|DG^n(x_0)w\|$$

we shall actually obtain an approximation of the largest LCE, call it $\lambda_1(x_0)$.

---

[2]The interested reader should consult Abarbanel *et al.* (1993).

In order to compute the smaller LCEs, we should select special initial vectors corresponding to the appropriate eigenvectors on the tangent space, which is impossible to do numerically. However, if we select *two* initial vectors w_1 and w_2 at random and apply $DG^n(x_0)$ to the unit-area parallelepiped formed by w_1 and w_2 (denoted by $\|w_1 \wedge w_2\|$), the asymptotic exponential rate of expansion (contraction) of the area of the parallelepiped will be

$$\lim_{n \to \infty} \frac{1}{n} \ln \|DG^n(x_0)w_1 \wedge DG^n(x_0)w_2\| = \lambda_1(x_0) + \lambda_2(x_0)$$

where $\lambda_2(x_0)$ is the second largest LCE. Having computed λ_1, λ_2 can thus be determined. The procedure can then be replicated for the remaining smaller LCEs, by considering sets of initial values of dimension $3, 4, \ldots, m$ and computing $\lambda_3, \ldots, \lambda_m$ by induction.

LCEs provide an extremely useful tool for characterising the behaviour of nonlinear dynamical systems. Since they are invariant with respect to topological conjugacy, it is sometimes possible to evaluate the LCEs by calculating those of a different and simpler map and establishing a conjugacy between the two maps. The signs of LCEs *calculated for a generic orbit* can be used to classify different types of attractors. The typical situations are summarised below.

(i) For asymptotically stable fixed points of an m-dimensional system, both in discrete and in continuous time, we shall have m negative LCEs. This should be clear from chapters 2 and 3, where it was seen that for stable fixed points, the distance between orbits beginning at different initial conditions in the basin of attraction, tends to decrease as orbits converge towards the equilibrium value.

(ii) In all other cases, continuous-time systems will have at least one LCE equal to zero. This corresponds to the direction defined by the vector tangent to the orbit at the relevant point.

(iii) For asymptotically stable limit cycles or quasiperiodic attractors of continuous-time m-dimensional systems we shall have $m - n$ negative LCEs ($n < m$) and n zero LCEs, with $n = 1$ for a limit cycle, or $n = k$ for quasiperiodic dynamics on a T^k torus. That is, in $m - n$ directions the distance between orbits decreases. For a limit cycle there is one direction (the one tangent to the periodic orbit), in which orbits neither approach nor diverge, for a quasiperiodic attractor there are k such directions (those spanning the space tangent to the T^k torus).

(iv) Chaotic attractors, both in discrete and in continuous-time, are associated with the presence of at least one positive LCE, which signals that nearby orbits diverge exponentially in the corresponding direction. It also means that observation errors will be amplified by the action of the map or the flow. As we see in chapter 9, the presence of one or more positive LCEs is intimately related to the lack of predictability of dynamical systems, which is an essential feature of chaotic behaviour.

7.2 Fractal dimension

The concept of (positive) Lyapunov exponents refers to a dynamic property of chaotic systems, i.e., the exponential divergence of nearby orbits. Chaotic sets, however, are typically characterised by a peculiar geometric structure usually denoted by the term 'fractal'. This name was derived from the Latin *fractus*, i.e., broken, by Benoît Mandelbrot (cf. 1982) to describe geometrical objects characterised by *self-similarity*, that is, objects having the same structure on all scales. Each part of a fractal object can thus be viewed as a reduced copy of the whole. Intuitively, a snowflake or a sponge can be taken as a natural fractal.

A simple, but precise mathematical example can be given as follows. Consider a segment of unit length; divide it into three equal subsegments and remove the intermediate one: you will be left with two segments of length 1/3. Divide each of them into three segments of length 1/9 and remove the (two) intermediate ones. If this procedure is iterated n times and we let n go to infinity, the 'surviving points', that is, the points that have not been discarded, form a set S called the **middle third Cantor set** (see figure 7.1). Let us then pose the following question: what is the dimension of S? Were n finite, S would be a collection of segments and its Euclidean dimension would be one. On the other hand, were S a finite collection of points, its Euclidean dimension would be zero. But in the limit for $n \to \infty$, S is an uncountable[3] collection of an infinite number of points forming nowhere an interval, and therefore the answer is not obvious. To deal with this problem, the traditional Euclidean notion of dimension is insufficient and we need the more sophisticated criterion of *fractal dimension*. There exists a rather large family of such dimensions, but we shall limit ourselves here to the simplest examples.

Let S be a set of points in a space of Euclidean dimension p. (Think,

---

[3]We say that a set S is **countable** if it is finite or it is **countably infinite** (i.e., if there is a one-to-one correspondence between S and the set of natural numbers). Sets which are not countable are called **uncountable**.

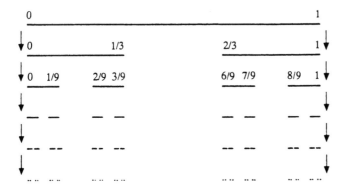

Fig. 7.1 The middle third Cantor set

for example, of the points on the real line generated in the construction of the Cantor set, or, more generally, by those resulting from the iterations of a one-dimensional map.) We now consider certain boxes of side ϵ (or, equivalently, certain spheres of diameter ϵ), and calculate the minimum number of such boxes, $N(\epsilon)$, necessary to cover S.[4]

Then, the fractal dimension D of the set S will be given by the following limit (assuming it exists):

$$D \equiv \lim_{\epsilon \to 0} \frac{\log[N(\epsilon)]}{\log(1/\epsilon)}. \tag{7.7}$$

The quantity defined in (7.7) is called the (Kolmogorov) **capacity dimension**. It is easily seen that, for the most familiar geometrical objects, it provides perfectly intuitive results. For example, if S consists of just one point, $N(\epsilon) = 1$ for any $\epsilon > 0$ and $D = 0$. If S is a segment of unit length, $N(\epsilon) = 1/\epsilon$ and $D = 1$. Finally, if S is a plane of unit area, $N(\epsilon) = 1/\epsilon^2$ and $D = 2$, etc. That is to say, for ordinary geometrical objects, the dimension D does not differ from the usual Euclidean dimension and, in particular, D is an integer. By making use of the notion of capacity dimension, we can now determine the dimension of the Cantor set constructed above. To evaluate the limit (7.7), we proceed step by step. Consider first a (one-dimensional) box of side ϵ. Clearly we shall have $N(\epsilon) = 1$ for $\epsilon = 1$, $N(\epsilon) = 2$ for $\epsilon = 1/3$, and, generalising, $N(\epsilon) = 2^n$ for $\epsilon = (1/3)^n$. Taking the limit for

---

[4]A technical definition of 'cover' is given in n. 1 on p. 108 of section 4.1. As a first approximation, here one can think of covering in the usual sense of the word.

$n \to \infty$ (or equivalently taking the limit for $\epsilon \to 0$), we can write

$$D = \lim_{\substack{n \to \infty \\ (\epsilon \to 0)}} \frac{\log 2^n}{\log 3^n} \approx 0.63.$$

We have thus provided a quantitative characterisation of a geometric set that is more complex than the usual Euclidean objects in that the dimension of S is noninteger. We might say that S is an object dimensionally greater than a point but smaller than a segment. It can also be verified that the Cantor set is characterised by self-similarity.

A second, well-known definition of fractal dimension, widely used in applications, is the **correlation dimension** (D_c). Strictly speaking, D_c is defined for an orbit rather than a set, but if the generating system is topologically transitive on a certain set (see definitions 4.8 and 4.9), the value obtained along a typical orbit also corresponds to the fractal dimension of the set.

Consider an orbit $S = \{x_0, x_1 = G(x_0), \ldots, x_N = G^N(x_0)\}$ generated by a map G on a subset $M \subset \mathbb{R}^m$. Consider all possible pairs (x_i, x_j) $i \neq j$, compute their (Euclidean) distance and verify whether $d(x_i, x_j) \lessgtr r$ for a given scalar value $r > 0$. Then write

$$C(r) = \lim_{N \to \infty} \frac{1}{N^2} \sum_{\substack{i,j=1 \\ i \neq j}}^{N} \theta \left[r - d(x_i, x_j) \right] \tag{7.8}$$

where θ is the Heavyside (counting) function

$$\theta(s) = \begin{cases} 1, & \text{if } s \geq 0; \\ 0, & \text{otherwise.} \end{cases}$$

The **correlation function** $C(r)$ gives the proportion of pairs of points whose distance is equal or less than $r > 0$, as a function of r, in the limit of an infinitely long sequence of points. If, for sufficiently small r, $C(r)$ can be approximated by r^{D_c}, where D_c is a constant, as numerical evidence seems to suggest (cf. Grassberger and Procaccia, 1983) then the **correlation dimension** of the orbit S is defined as

$$D_c = \lim_{r \to 0} \frac{\ln C(r)}{\ln r}.$$

In practice, only not too small and not too large values of r can be considered. If the orbit is bounded, for sufficiently large values of r, $C(r)$ tends to one (due to the saturation effect, all the points are counted). On the other hand, when the number of points generated by the map G is finite, for sufficiently small r, $C(r)$ tends to zero (no point is counted). In numerical estimates, in which the number of points is limited and precision is finite, for too small r the value is estimated poorly, since the number of observations is small and errors become important. In practice, one tries to define a *scaling region* with lower and upper bounds (in terms of r) over which the slope of the $\log C(r)$ versus $\log r$ curve is approximately constant.

A third definition of fractal dimension directly relates the dimension of an attractor and the associated LCEs suggested by Kaplan and Yorke (1979). Consider the typical orbit on a chaotic attractor of a dynamical system (a flow or a map) and the associated set of Lyapunov exponents ordered from the largest λ_1 to the smallest λ_m. Suppose that the j is the largest integer for which $\sum_{i=1}^{j} \lambda_i > 0$ (this implies that $\lambda_{j+1} < 0$). Then the dimension of the attractor is calculated as

$$D_L = j + \frac{\sum_{i=1}^{j} \lambda_i}{|\lambda_{j+1}|}$$

and called **Lyapunov dimension**. There exists some numerical evidence that D_L gives a good numerical approximation of D and D_c (see Abarbanel *et al.*, 1993 and the bibliography therein). For the Lorenz attractor like that presented in figure 6.10, estimates of the LCEs are $\lambda_1 = 0.905 \pm 0.005$, $\lambda_2 = 0.0$, $\lambda_3 = -14.57 \pm 0.01$, giving the Lyapunov dimension $D_L = 2 + 0.905/14.57 \approx 2.062$ (see, for example, Alligood *et al.*, 1997, p. 366).

The Kolmogorov capacity and the correlation dimension are prototypical members of a family of dimension-like quantities which can be grouped under the label *fractal dimensions* and are suitable for fractal objects, i.e., objects characterised by a noninteger dimension. When applied to the geometrical analysis of dynamical systems, the fractal dimension can be taken as a measure of the way orbits fill the state space under the action of a flow (or a map). For example, we know that some noninvertible, one-dimensional maps, for certain values of the controlling parameter, are characterised by invariant sets (which may or may not be attracting) which are Cantor sets.[5] Two-dimensional maps may have invariant sets of the horseshoe type which, as we saw in section 6.4, are the products of Cantor sets.

The association between noninteger fractal dimension and chaos for dissipative, continuous-time dynamical systems can be understood by considering that their complex dynamics are the result of the infinitely repeated stretching and folding of a bundle of orbits under the action of the flow. To visualise this idea in a nonrigorous, but intuitively appealing manner, let

---

[5]We saw an example of a Cantor set generated by the logistic map in chapter 6, section 6.3 and we shall see another example in chapter 8 called the Feigenbaum attractor.

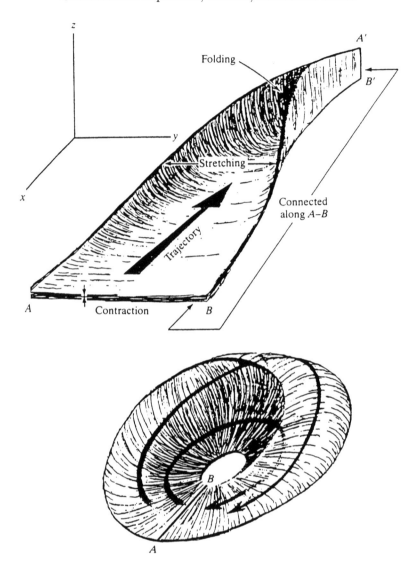

Fig. 7.2 Stretching and folding in a chaotic attractor. Reproduced by permission of *Zeitschrift für Naturforschung* from Shaw (1981)

us consider a flow acting on a bounded subset of a three-dimensional state space constructed in the following way (see figure 7.2).

Orbits of the flow move largely in one direction (indicated by the big arrow); they tend to spread out in a second direction (intermediate arrow) and they are pushed together in a third one (small arrows). As a result of this compression, a three-dimensional volume of initial conditions will be squeezed onto a (quasi) two-dimensional *sheet*. But, although nearby orbits diverge exponentially, the motion of the system, by construction, is bounded. Therefore the sheet on which the flow takes place, though continuously expanded along its width, must be folded over on itself and, for the same reason, the two ends AB, $A'B'$ of the sheet must be twisted and smoothly joined together. However, since $A'B'$ has two distinct sheets and, at each turn, it joins AB which has only one, in order for the joining to be smooth (which is a necessary condition for the system to be representable by a flow), the sheet on which the flow occurs must in fact be a 'book' of thickness between 0 and 1, made of infinitely many 'leaves', each of zero thickness. In fact, if we position a two-dimensional Poincaré surface of section across the orbits, the resulting set of intersection points will have a fractal structure similar to that generated by the horseshoe map. When the compression (small arrows) is very strong, the resulting Poincaré map may look one-dimensional.[6]

A mechanism broadly corresponding to the one just described seems also to be at the root of the complex dynamics of certain three-dimensional dynamical systems investigated in the literature such as the so-called Rössler attractor (1976), see figure 7.3 for two perspectives. The geometrical aspect of the attractors of these systems, as well as the relevant numerical calculations, indicate that their fractal dimensions are noninteger and between 2 and 3.

In fact, any chaotic attractor of a continuous-time dissipative system in $\mathbb{R}^3$ must have a fractal dimension $2 < D < 3$. To see this, let us consider that, because of dissipation, the volume of any initial three-dimensional set of initial conditions asymptotically shrinks under the action of the flow. This means that the fractal dimension D of any attractor of the system must be smaller than 3. On the other hand, for the attractor to be chaotic (in the sense that it possesses SDIC), its dimension must be greater than 2: no attractor more complex than a limit cycle may occur in $\mathbb{R}^2$. Consequently, it must be the case that $2 < D < 3$.

remark 7.4 Invariant sets or attractors, characterised by noninteger fractal dimension, are often called 'strange'. Notice, however, that 'chaoticity', as defined by SDIC, and 'strangeness' are independent properties, at least for discrete-time systems. Thus, we have chaotic attractors that are not fractal and strange (fractal) attractors that are not chaotic. A well-known example of the second case is the so-called *Feigenbaum attractor* which we encounter in our discussion of the period-doubling bifurcation in chapter 8.

---

[6]See, on this point, Cvitanović (1984), p. 18.

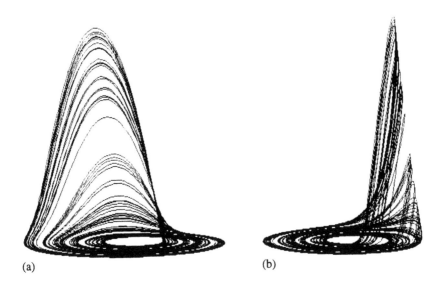

(a)   (b)

Fig. 7.3 The Rössler attractor

An equally well-known example of the first case is the attractor[7] generated by the logistic map (with parameter $\mu = 4$) or the tent map. For continuous-time dynamical systems, however, it is conjectured that, in general, strange attractors are also chaotic.

The concept and measurement of fractal dimension are not only necessary to understand the finer geometrical nature of strange attractors, but they are also very useful tools for providing quantitative analyses of such attractors, in both theoretical and applied investigations of dynamical systems. A well-known example of application of the notion of correlation dimension to the analysis of time series is the so-called BDS test (see Brock *et al.*, 1996).

7.3 Horseshoes and homoclinic orbits

In section 6.4 we analysed the prototypical horseshoe model and showed how the combination of stretching in one direction and contracting in the other may generate chaotic behaviour (although not necessarily a chaotic attractor). In his seminal work Smale (1967) established a general relation between horseshoe-like dynamics and homoclinic points, proving that the existence of a *homoclinic point* of a diffeomorphism implies the presence of *horseshoe dynamics* and therefore of an invariant chaotic set. Homoclinic

---

[7]Here, it is an attractor in the weak sense of definition 4.11.

points (and homoclinic orbits) play a fundamental role in the generation of chaos and we shall discuss them in some detail.

7.3.1 Homoclinic orbits for maps

Consider first the case of a diffeomorphism G on $\mathbb{R}^2$, which may or may not be a Poincaré map derived from a flow in $\mathbb{R}^3$. Suppose the system has a hyperbolic fixed point p of a saddle type, and that $W^s(p)$ and $W^u(p)$ are its stable and unstable manifolds.[8] Suppose now that $W^s(p)$ and $W^u(p)$ intersect transversally (see definition below) at a point $q \in \mathbb{R}^2$ (see figure 7.4(a)).

Notice that the curves W^u and W^s are *not* orbits of a flow which, by uniqueness of solutions of differential equations, could not cross. They are instead one-dimensional subsets of the state space of a map, each containing infinitely many orbits $\{G^i(x_0)\}_{i=0}^k$ which asymptotically tend to p as k tends to $+\infty$ if $x_0 \in W^s$, or as k tends to $-\infty$ if $x_0 \in W^u$. We now need the following definitions.

definition 7.1 *Let $G : M \to M$ be a diffeomorphism of a metric space M. A point $x \in M$ is* **homoclinic** *to $y \in M$ if*

$$\lim_{|n| \to \infty} d\left[G^n(x), G^n(y)\right] = 0$$

where d is a distance function on M. The point x is said to be **heteroclinic** *to the points $y_1, y_2 \in M$ if*

$$\lim_{n \to +\infty} d\left[G^n(x), G^n(y_1)\right] = 0$$

and

$$\lim_{n \to -\infty} d\left[G^n(x), G^n(y_2)\right] = 0.$$

Notice that if y, y_1, y_2 are fixed points, $G^n(y), G^n(y_1), G^n(y_2)$ can be replaced by y, y_1, y_2, respectively.

definition 7.2 *If p is a fixed point of G, we say that $q \in M$ is a* **homoclinic point** *of p if $q \in [W^s(p) \cap W^u(p)]$. If the intersection of $W^s(p)$ and $W^u(p)$ is transversal, we say that q is a transversal homoclinic point.*

Transversality is a subtle concept and we shall discuss it only in a very

---

[8]The analysis can be generalised to the case in which p is a periodic point of G of period k. In this case we can consider any of the k fixed points of the map G^k.

concise manner, referring the reader to the specialised literature. What follows is based on Wiggins (1990), pp. 101–2. (For a more detailed treatment see Robinson, 1999, pp. 453–5 and the bibliography provided there.)

definition 7.3 *Let M and N be differentiable manifolds in $\mathbb{R}^m$ and let q be a point in $\mathbb{R}^m$. We say that M and N are **transversal** at q if $q \notin M \cap N$ or, if $q \in (M \cap N)$, then span $(T_q M \cup T_q N) = \mathbb{R}^m$ (i.e., the tangent spaces at q of M and N, respectively $T_q M$ and $T_q N$, span or generate the space $\mathbb{R}^m$). We simply say that M and N are transversal if they are transversal at all points $q \in \mathbb{R}^m$.*

According to definitions 7.1–7.3 because W^s and W^u are invariant, all forward and backward iterates of the map G must lie both on W^s and on W^u, that is, $G^k(q)$ or $G^{-k}(q)$ are also homoclinic points, for all $k > 1$. Thus, if the stable and unstable manifolds of a fixed point intersect once, they must intersect a (doubly) infinite number of times. This generates wildly oscillating loops that intersect one another in an extremely complicated manner, giving rise to a so-called **homoclinic tangle**, see figure 7.4(a). A detailed description of the amazingly intricate geometry of the tangle can be found in, for example, Wiggins (1990), pp. 519–40.

remark 7.5 Similarly complex situations arise when the stable manifold of a hyperbolic fixed point p_1 of G intersects transversally the unstable manifold of another fixed point p_2 *and* the unstable manifold of p_1 intersects transversally the stable manifold of p_2. These are called **heteroclinic (transversal) intersections** and the corresponding **heteroclinic tangle** is summarily sketched in figure 7.4(b).

Smale's horseshoe construction can be regarded as an idealised mathematical description of the dynamics of a system near a homoclinic point. The connection between transversal homoclinic points and horseshoes is by no means limited to diffeomorphisms of the plane, but can be stated in a general form as follows (see Katok and Hasselblatt, 1995, p. 276; Robinson, 1999, p. 288).

theorem 7.1 (Smale–Birkhoff) *Suppose q is a transversal homoclinic point of a hyperbolic periodic point p for a diffeomorphism G. Then, for each neighbourhood N of $\{p, q\}$ there exists a positive integer n such that the map G^n has a hyperbolic invariant set $\Lambda \subset N$ with $p, q \in \Lambda$ and on which G^n is topologically conjugate to the two-sided shift map T on a two symbol space Σ_2.*

We know that the dynamics of T on Σ_2 is chaotic and so, therefore, is the dynamics of G^n on Λ.

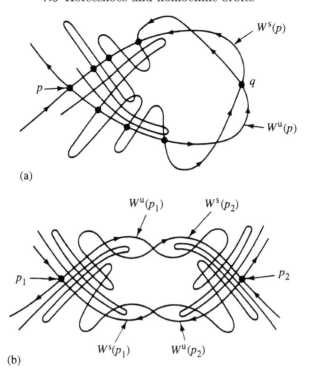

(a)

(b)

Fig. 7.4 (a) Homoclinic tangle; (b) heteroclinic tangles. Reproduced by permission of Cambridge University Press from Ott, *Chaos in Dynamical Systems* (1993)

The Smale–Birkhoff theorem is one of the few rigorous general results in chaos theory providing a precise *sufficient* condition for the existence of chaotic dynamics (although not necessarily for the existence of a chaotic *attractor*). Some authors even define chaotic (or strange) dynamics in terms of the presence of transversal homoclinic orbits (see Guckenheimer and Holmes, 1983, p. 256). However, we do not know of a converse result for theorem 7.1.

remark 7.6 Notice that, whereas the stable manifold of a fixed point of an invertible map can intersect its unstable manifold, stable or unstable manifolds cannot self-intersect. The proof of this statement is left to the reader as an exercise (see Ott, 1993, pp. 122–3).

7.3.2 Homoclinic points for flows

Notice, first of all, that stable and unstable manifolds of a fixed point of continuous-time flows in $\mathbb{R}^m$ *cannot* intersect transversally. To see this, consider that for two vector subspaces intersecting in $\mathbb{R}^m$, it must be

$$\dim[\text{span}(T_q M \cup T_q N)] = \dim(T_q M) + \dim(T_q N) - \dim(T_q M \cap T_q N).$$

On the other hand, according to definition 7.3, if M and N intersect transversally, it follows that

$$m = \dim(T_q M) + \dim(T_q N) - \dim(T_q M \cap T_q N).$$

Now, by uniqueness of solutions of differential equations, if the stable and unstable manifolds of a fixed point p intersect at a point q, they must intersect along the (unique) orbit passing through q (see figure 7.5(a)). Setting $W^s(p) = M$ and $W^u(p) = N$, this implies

$$\dim(T_q M \cap T_q N) \geq 1.$$

But for a hyperbolic fixed point $p \in \mathbb{R}^m$ we have

$$\dim(T_q M) + \dim(T_q N) = \dim(\mathbb{R}^m) = m.$$

Thus transversal intersections of $W^s(p)$ and $W^u(p)$ are impossible.

However, for a hyperbolic *periodic solution* Γ of a flow on $\mathbb{R}^m$, we have, for any point $q \in \Gamma$

$$\dim(T_q M) + \dim(T_q N) = m + 1$$

and transversal intersections are possible. Figure 7.5(b) gives a (partial) representation of the case in which stable and unstable manifolds of a periodic orbit Γ, $W^s(\Gamma), W^u(\Gamma)$, respectively, intersect transversally in $\mathbb{R}^3$ along the orbit γ. The presence of a transversal homoclinic point to a periodic orbit of a flow induces a horseshoe-like invariant set. In fact, under certain conditions chaotic invariant sets of a horseshoe type can be brought about even by *nontransversal* homoclinic orbits of fixed points of flows. An example of this was observed in the Lorenz model discussed above. Another example is provided by a model of three differential equations investigated by Shilnikov and whose (chaotic) dynamics is known as the 'Shilnikov phenomenon'. (For details see Guckenheimer and Holmes, 1983, p. 50, pp. 318–25; Robinson, 1999, pp. 299–302.) A sketch of nontransversal intersections of stable and unstable manifolds of a fixed point p for a flow in $\mathbb{R}^3$ is presented in figure 7.5(c).

7.3.3 Horseshoes in the Hénon map

Hénon's is one of the canonical 'toy maps' specifically conceived to investigate chaotic dynamics by means of iterations of simple functions easily performed on modern computers. Like the Lorenz model, from which it

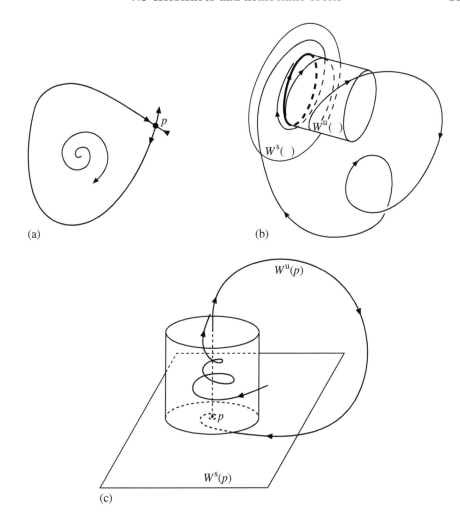

(a)   (b)

(c)

Fig. 7.5 (a) Homoclinic orbit in $\mathbb{R}^2$; (b) transversal intersection of stable and unstable manifolds of a periodic orbit in $\mathbb{R}^3$. Reproduced by kind permission from Kluwer Academic Publishers from Neimark and Landa, *Stochastic and Chaotic Oscillations* (1992), figure 4.3; (c) nontransversal intersection of stable and unstable manifolds of a fixed point in $\mathbb{R}^3$

was derived, the Hénon map has been extensively studied for more than two decades and has inspired many interesting theoretical results.

The map has the form $z_{n+1} = G(z_n)$ where $z \in \mathbb{R}^2$, $z_n = (x_n, y_n)$, $x, y \in \mathbb{R}$, specifically:

$$x_{n+1} = a - x_n^2 + by_n$$
$$y_{n+1} = x_n$$

Fig. 7.6 The Hénon map

and a typical trajectory for the chaotic parameter configuration is provided in figure 7.6.

It has been established that for a sufficiently large, precisely, for $a \geq (5 + 2\sqrt{5})(1 + |b|)^2/4$, $b = \pm 0.3$, there exists an invariant (Cantor) set Λ in the plane such that the dynamics of G on Λ (denoted $G|\Lambda$) is topologically conjugate to the shift map T on the two-symbol space Σ_2, thus $G|\Lambda$ is chaotic. Proof of this result can be found in Robinson (1999), pp. 283–7.

Whether the Hénon map can generate a chaotic *attractor* has been an open question for a very long time. Computer simulations for the canonical values of parameters, $a = 1.4$ and $b = 0.3$ suggested that this is the case. Finally, a result by Bendicks and Carleson (1991) established that for very small values of b, there indeed exists an attractor with a positive Lyapunov exponent. Whether this is also true for $b = 0.3$ is still undecided.

remark 7.7 The notion of horseshoes is not restricted to diffeomorphisms of the plane, but can be generalised to diffeomorphisms of $\mathbb{R}^m$. A precise definition of such *general horseshoes* can be found in Katok and Hasselblatt (1995), p. 274; see also Robinson (1999), pp. 287–96.

Exercises

7.1   Use the procedure explained in section 7.1 to estimate the Lyapunov characteristic exponents of a generic orbit from initial value x_0 for the following maps

$$G_D : [0, 1) \to [0, 1) \qquad G_D(x) = 2x \bmod 1$$
$$G : \mathbb{R} \to \mathbb{R} \qquad\qquad G(x) = 2x.$$

What can be said about the generic dynamics of these maps on the basis of these LCEs?

7.2   Consider the logistic map G with $\mu = 3.2$.

(a) Verify that the generic orbit converges to a periodic orbit of period 2. Evaluate its LCE.

(b) The LCE is related to the propagation of initial errors. Let e_0 be the initial error in measuring x_0, then $e_n = G^n(x_0 + e_0) - G^n(x_0)$ is the error after n iterations and, for n sufficiently large, we expect

$$\lambda(x_0) \approx \frac{1}{n} \ln \left| \frac{e_n}{e_0} \right|.$$

Suppose $x_0 = 0.3$ but the measuring device gives $\hat{x}_0 = 0.31$. Calculate the approximate LCE after a number of iterations and check convergence towards the known exact value.

7.3   Repeat part (b) of the preceding exercise for $\mu = 4$.

7.4   Consider the map of the T^2 torus (additive notation)

$$G : S^1 \times S^1 \to S^1 \times S^1$$
$$G\begin{pmatrix} x \\ y \end{pmatrix} = \begin{pmatrix} 1 + x & \bmod 1 \\ \sqrt{2} + y & \bmod 1 \end{pmatrix}.$$

(a) Evaluate the LCEs of a typical orbit of the map;

(b) Prove that G is topologically transitive on $S^1 \times S^1$.

7.5   Consider the Lorenz model described by (6.7). Verify that figure 6.13 is a plot of successive maxima of $x(t)$, for values of parameters $\sigma = 10$, $b = \frac{8}{3}$, $r = 28$. The plot can be approximated by the map

$$G : [0, 1] \to [0, 1]$$
$$G(x) = \begin{cases} \frac{(2-a)x}{1-ax}, & \text{if } 0 \le x \le \frac{1}{2}; \\ \frac{(2-a)(1-x)}{1-a(1-x)}, & \text{if } \frac{1}{2} < x \le 1 \end{cases}$$

where $a = 0.99$. Verify that for a typical orbit of G, the LCE > 0.

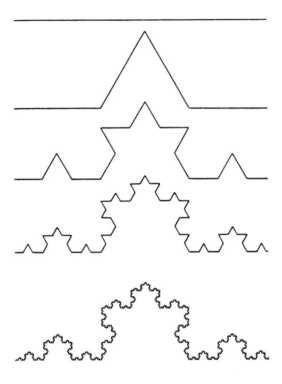

Fig. 7.7 One side of the Koch snowflake

7.6   Construct the so-called 'Koch snowflake' (see figure 7.7) starting
      with an equilateral triangle with sides of unit length and then re-
      peating this algorithm 3 or 4 times: remove the middle third of each
      side and replace it with two pieces of equal length arranged as a tent.
      After the first iteration the triangle should have become a 6-pointed
      star with 12 sides, each of length 1/3. At each subsequent iteration
      the star shape gets more delicate, the sides of length 1/9, 1/27, etc.
      Calculate the capacity dimension (7.7) for the Koch snowflake.

7.7   Let N_i denote the number of sides of the Koch snowflake at the ith
      iteration and L_i denote the length of one side after the ith iteration.
      Write the formula for calculating the perimeter P_i of the snowflake
      after i iterations and show that for $i \to \infty$, $P_i \to \infty$.

7.8   The fractal set known as the Sierpinski triangle is constructed by
      removing pieces, as for the Cantor set. Begin with an equilateral
      triangle of unit sides. Join the midpoints of each side and remove
      (blacken) the middle triangle. In the 3 remaining equilateral trian-
      gles, again join the midpoints and remove the middle triangle. Re-

peat this process for the 9 remaining triangles, 27 triangles, etc. The limit set (as the number of iterations tends to infinity) is denoted by T, the Sierpinski triangle. Let N_i be the number of triangles necessary to cover T after the ith iteration, L_i be the length of the side of the corresponding triangle. Calculate the fractal dimension of the Sierpinski triangle.

7.9   Why, in computing the proportion in (7.8), do we divide by N^2?

7.10   Find an example of a Cantor subset of the interval with positive fractal dimension but zero length.

8

Transition to chaos

In chapter 5 we studied bifurcations, that is, qualitative changes in the orbit structures of dynamical systems, which take place when parameters are varied. In chapter 6 we discussed chaos and provided a precise characterisation of chaotic dynamics. In this chapter we take up again the question of transition in a system's behaviour, with a view to understanding how complex dynamics and chaos appear as parameters change. This problem is often discussed under the label 'routes to chaos'. The present state of the art does not permit us to define the prerequisites of chaotic behaviour with sufficient precision and generality, and we do not have a complete and exhaustive list of all such possible routes. In what follows we limit our investigation to a small number of 'canonical' transitions to chaos which are most commonly encountered in applications. In our discussion we omit many technical details and refer the reader to the relevant references.

8.1 Period-doubling route to chaos

Although the period-doubling route to chaos could be discussed in a rather general framework (cf. Eckmann 1981, pp. 648–9), here we shall treat it in the context of noninvertible one-dimensional maps, because they provide an interesting topic *per se* and are by far the most common type of dynamical system encountered in applications. Before describing the period-doubling scenario in detail, we discuss some general results covering a broad class of one-dimensional maps.

8.1.1 Maps of the interval

In this section we work with one-parameter families of smooth maps of the interval, $G_\mu : [0,1] \to [0,1]$, $\mu \in I_\mu \subset \mathbb{R}$, belonging to the **unimodal** class, that is, such that for all $\mu \in I_\mu$

(i) $G_\mu(0) = G_\mu(1) = 0$;
(ii) G_μ has a unique critical point $x_c \in (0,1)$.

definition 8.1 *A point x_c is said to be a **critical point** if $G'(x_c) = 0$. The critical point is, respectively, nondegenerate or degenerate, according to whether $G''(x_c) \neq 0$, or $G''(x_c) = 0$ (where $G'(x) = dG/dx$ and $G''(x) = d^2G/dx^2$).*

Unimodal maps are quite well understood and we include here some fundamental results which are independent of the specific form they take in different applications.

Suppose G is three times differentiable, and either of the following equivalent properties holds:

$\sqrt{|G'|}$ is a convex function or;

$$S(G) \equiv \frac{G'''(x)}{G'(x)} - \frac{3}{2}\left(\frac{G''(x)}{G'(x)}\right)^2 < 0 \quad \text{where} \quad G'''(x) = \frac{d^3G}{dx^3}$$

and $S(G)$ is usually called the **Schwarzian derivative**.

For unimodal maps of the interval with negative $S(G)$ the following properties hold.

(i)   For any given value of the parameter μ, orbits starting from almost all initial points on the interval $[0,1]$ have the same ω-limit set Λ and therefore the same asymptotic behaviour ('almost' here means excluding a set of Lebesgue measure zero). This result is important because it means that the choice of initial conditions is not important as far as the asymptotic behaviour of the map is concerned.
(ii) The simplest case obtains when Λ is a finite set of points and the dynamics is periodic of period n. Unimodal maps with $S(G) < 0$ can have *at most* one *stable* periodic orbit of period $n \geq 1$ ($n = 1$ corresponds to fixed points) and the orbit of the unique critical point is asymptotic to this stable orbit.
(iii) When the set Λ is not a periodic orbit, we have the following two possibilities. 1. The limit set Λ is a Cantor set on which the generic dynamics are aperiodic but not chaotic, and the map G on Λ does

not have sensitive dependence on initial conditions. (We shall see in chapter 9 that in the aperiodic nonchaotic case, the metric entropy of the map is zero.) 2. The set Λ is a finite union of n intervals $(n \geq 1)$, containing the critical point in its interior. Λ is bounded by the first $2n$ iterates of the critical point. The dynamics on Λ is chaotic in the sense that the map G has SDIC on Λ (also, as we shall see, the metric entropy of G is positive). The reader can verify that for the logistic map with $\mu > 3.68$ (approximately), $n = 1$ and at $\mu = 4$, Λ consists of a unique interval equal to the entire state space $[0, 1]$.

remark 8.1 The first part of the result in (ii) (uniqueness of the stable periodic orbit) is important when we face the problem of distinguishing truly chaotic from noisy periodic behaviour. For unimodal maps with a negative Schwarzian derivative we can never have a multiplicity of disconnected, stable periodic orbits, each with a very small basin of attraction. Were this the case, small perturbations, physical or numerical, would prevent orbits from converging to their (periodic) asymptotic behaviour, suggesting the presence of chaos where there is none. Some authors have hinted that this might be the case for the (two-dimensional) Hénon map with 'canonical' parameter values $a = 1.4$, $b = 0.3$. As we mentioned at the end of the last chapter, in section 7.3, although the existence of 'true chaos' has been proved for another set of parameter values, a definite answer to the question for the canonical parameter values is not yet available.

8.1.2 Period 3

Period-3 orbits are of particular importance for maps of the real line. A very general result and some corollaries will illustrate the point. Consider the following ordering of the natural numbers (the expression $p \succ q$ simply means 'p is listed before q'):

$$3 \succ 5 \succ 7 \succ \cdots \succ 2 \cdot 3 \succ 2 \cdot 5 \succ \cdots \succ 2^2 \cdot 3 \succ 2^2 \cdot 5 \succ \cdots$$

$$\succ 2^n \cdot 3 \succ 2^n \cdot 5 \succ \cdots \succ 2^n \succ \cdots \succ 2^3 \succ 2^2 \succ 2 \succ 1.$$

In other words, we have first all the odd numbers greater than one; then 2 times the odd numbers, 2^2 times the odd numbers, 2^3 times the odd numbers, etc. What remains of the natural numbers are only the powers of two which are listed in decreasing order $(1 = 2^0)$.

theorem 8.1 (Sharkowskii) *Let $G : I \rightarrow I$ with $I \subseteq \mathbb{R}$ be a continuous map. If G has a periodic point of prime period k, and $k \succ l$ in the above ordering, then G also has a periodic point of period l.*

Sharkowskii's theorem is very general and applies to all *continuous* maps of $\mathbb{R}$. On the other hand, it is a strictly one-dimensional result and it does not hold for higher-dimensional maps, in fact not even for maps on the circle.

Within its limits, this theorem is a very powerful (and aesthetically beautiful) tool of investigation. Two implications of the theorem are particularly interesting for us here. First, if G has a periodic point whose period is *not* a power of 2, then G must have infinitely many periodic orbits. Conversely, if there exists finitely many periodic orbits, they must necessarily be of period 2^k ($k \geq 1$). Moreover, theorem 8.1 implies that if G has an orbit of period 3, it must also have uncountably many orbits of infinite period, that is, aperiodic orbits.

Li and Yorke (1975) independently proved the main results of Sharkowskii's theorem and argued that 'period three implies chaos' in the sense that if G has a period-3 orbit, then there exists an uncountable set S of aperiodic orbits (sometimes called a 'scrambled set') which is chaotic in the sense that:

(i) an arbitrary point $x \in S$ and any periodic point $\bar{x}$ of G satisfy

$$\lim_{n \to \infty} \sup |G^n(x) - G^n(\bar{x})| > 0;$$

(ii) any two arbitrary points $x, y \in S$ satisfy

$$\lim_{n \to \infty} \sup |G^n(x) - G^n(y)| > 0;$$
$$\lim_{n \to \infty} \inf |G^n(x) - G^n(y)| = 0.$$

Thus, orbits starting in S do not asymptotically converge to any periodic orbit. Also every two orbits starting in S come arbitrarily close to one another but do not approach one another asymptotically. In this sense, the dynamics of G on S are chaotic. However, period 3 does not imply that G possesses a chaotic *attractor*. This is most conspicuously illustrated by the behaviour of the logistic map inside the period-3 window. Here, for almost any initial condition in $[0, 1]$ (in terms of Lebesgue measure) the orbits of G converge asymptotically to the unique stable periodic orbit of period 3. Chaotic behaviour (in the sense described above) obtains only for a set of initial conditions of measure zero. The presence of a nonattracting scrambled set, however, does affect the transient dynamics for a more or less long period of time.

8.1.3 The logistic map

A prototypical example of a unimodal map with SDIC for some parameter
values is provided by the logistic map

$$x_{n+1} = \mu x_n (1 - x_n) \qquad \mu \in [1, 4].$$

This map was used in particular in section 5.4 to study the flip bifurcation and we resume our discussion where we left off. Recall that the map
has fixed-point branches defined by $\bar{x}_1 = 0$ and $\bar{x}_2 = (\mu - 1)/\mu$ and the
following bifurcation scenario. There is a transcritical bifurcation, with the
accompanying exchange of stability at $\mu = 1$ (with $\bar{x}_1 = \bar{x}_2 = 0$). The first
flip bifurcation occurs at $\mu = 3$, leading to a stable period-2 cycle which
eventually loses stability at $\mu = 1 + \sqrt{6} \approx 3.45$ and gives rise to a stable
period-4 cycle. As μ increases further, the scenario repeats itself over and
over again: each time a period-2^k cycle of the map G loses stability through
a flip bifurcation of the map G^{k+1} which gives rise to an initially stable
period-2^{k+1} cycle, and so on and so forth. The sequence $\{\mu_k\}$ of values
of μ at which cycles of period 2^k appear has a finite accumulation point
$\mu_\infty \approx 3.57$. At $\mu = \mu_\infty$ there exists an infinite number of periodic orbits
with periods equal to powers of 2, all of them unstable. At that value of
the parameter μ, for almost all initial points, the ω-limit set Λ is known
as the **Feigenbaum attractor**. It is an infinite set of points (actually a
Cantor set), which is strange in the sense that it has a fractal dimension
$(D(\Lambda) \approx 0.54)$. However, for a generic orbit, the single Lyapunov characteristic exponent corresponding to μ_∞ is null, there is no sensitive dependence
on initial conditions and the dynamics are not chaotic. Moreover, although
Lyapunov stable in the sense of definition 4.5, the Feigenbaum attractor is
not asymptotically stable in the sense of definition 4.6 and, therefore, is not
an attractor in the sense of definition 4.10. It is instead an attractor in the
weaker sense of definition 4.11.

Feigenbaum (1978) discovered, and Collet *et al.* (1980) and Lanford
(1982) later rigorously proved, that there are two universal quantitative
features in the period-doubling scenario:

(i) The convergence of μ to μ_∞ is controlled by the universal parameter
   $\delta \approx 4.67$. If we call Δ_i and Δ_{i+1}, respectively, the distances on the
   real line between successive bifurcation values of the parameter (i.e.,
   $\Delta_i = \mu_k - \mu_{k-1}$ and $\Delta_{i+1} = \mu_{k+1} - \mu_k$, $k \geq 2$), we have:

$$\lim_{i \to \infty} \frac{\Delta_i}{\Delta_{i+1}} = \delta.$$

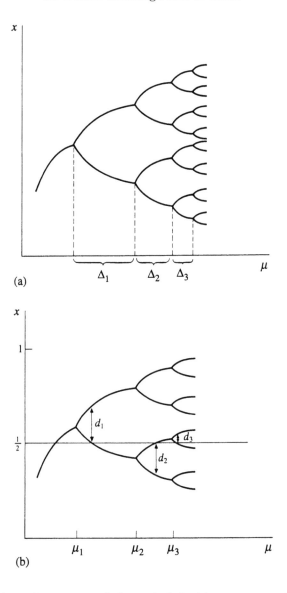

Fig. 8.1 The universal constants of the period-doubling scenario. Reproduced by permission of Wiley-VCH from Schuster, *Deterministic Chaos: An Introduction*, 2nd edn (1989)

The number δ has become known as the **Feigenbaum constant** (see figure 8.1(a)).

(ii) The relative scale of successive **branch splittings** measured by the distance between the two branches of periodic points corresponding to a superstable k-cycle[1] is controlled by the universal parameter $\alpha \approx 2.50$

[1]A k-cycle is **superstable** if the derivative of the iterated map G^k at a periodic point is zero,

(see figure 8.1(b)). For a superstable cycle, the critical point $x = 1/2$ must be one of the periodic points. If $x_k \neq 1/2$ is the periodic point closest to $1/2$ and $d_i = (x_k - 1/2)$, with $k = 2^i$ and $(i = 1, 2, \ldots)$ then

$$\lim_{i \to \infty} \frac{d_i}{d_{i+1}} = -\alpha.$$

The parameters δ and α are universal in the sense that, for a very large class of maps, they have the same values. These numbers also agree with numerical calculations performed in a variety of physical experiments. There is some evidence, moreover, that this universality applies not only to simple one-dimensional maps, but also to continuous-time, multi-dimensional systems which, owing to strong dissipation, can be characterised by a one-dimensional map.[2] For practical purposes, universality often permits us to make accurate predictions of μ_∞ as soon as the first few bifurcation values are known. For example, in the case of the logistic map, the knowledge of μ_2 and μ_3 allows us (using a pocket calculator) to predict μ_∞ correctly to four decimal places.

In figure 8.2(a) the upper part provides the bifurcation diagram whereas the lower part provides the associated single LCEs for a typical orbit of the logistic map over $2.8 \leq \mu \leq 4$. Note that at the bifurcation points, as well as at μ_∞, the LCEs are equal to zero.

For $\mu_\infty < \mu \leq 4$, we are in the chaotic zone or chaotic regime, where, to each value of the controlling parameter μ there corresponds one of the three basic types of dynamics (periodic, aperiodic nonchaotic, chaotic) described above for the general unimodal class. Turning again to figure 8.2(a) notice that, because of approximation, LCEs look mostly positive (thus the name **chaotic zone**), but there are many downward spikes corresponding to periodic attractors. In applications, however, one must assume the presence of a certain amount of noise and the practical relevance of most periodic windows may be questioned. In figure 8.2(b), the map has been perturbed at each iteration by some small additive noise. It can be seen that the addition of noise washes out the finer structure in the bifurcation of the LCE diagrams. There is still a sharp transition to the chaotic zone, but only

namely:

$$\frac{dG^k(x_0)}{dx} = \prod_i G'(x_i) = 0, \qquad (i = 0, 1, \ldots, k-1).$$

In the case of a unimodal map, this implies that a superstable cycle always contains the critical point because it is the only point at which $G'(x) = 0$.

[2] Cf. Franceschini and Tibaldi (1979), in which infinite sequences of period-doublings have been found in a system of five differential equations. The numerically-calculated values of δ and α agree with the Feigenbaum constants.

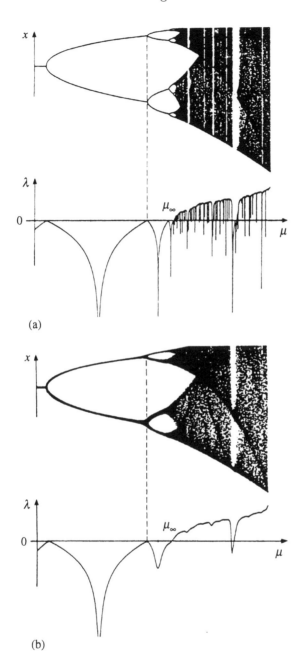

Fig. 8.2 Bifurcation diagrams and Lyapunov characteristic exponents for the logistic map: (a) $2.8 \leq \mu \leq 4$; (b) same with additive noise. Reproduced by permission of Wiley-VCH from Schuster, *Deterministic Chaos: An Introduction*, 2nd edn (1989)

the largest of periodic windows remain visible. Which periodic windows are washed out depends on the type and support of the noise process (on this point see, for example, Schuster, 1989, pp. 59–62).

The limit case $\mu = 4$ deserves special attention. As we showed in section 6.5, for this value of the controlling parameter, the logistic map is topologically conjugate to the tent map, and is therefore chaotic (it has SDIC on $[0, 1]$).

The route to chaos we have been discussing is probably the most common one in applications. Notice that the asymptotic orbit structure evolves gradually through a sequence of continuous bifurcations, brought about by variations of the control parameter. Consequently, there are plenty of forewarnings of the impending chaotic behaviour (hence the name of **safe-boundary bifurcation** used in the Russian literature for this kind of scenario). As we shall shortly see, however, the smooth, period-doubling route to chaos is not the only one existing for one-dimensional maps of the logistic type.

8.2 Intermittency

Generally speaking, a signal (the value of a variable changing in time) is said to be intermittent if it is subject to infrequent variations of large amplitude. In the present context, we apply the term **intermittency** to a phenomenon, fairly common in low-dimensional dynamical systems in continuous or discrete time depending on a parameter, whose main features can be given the following broad description. Consider a map G_μ depending on a parameter μ. For values of μ less than[3] a certain critical value μ_c, the typical orbits of the system converge to an invariant set A_1 with a simple structure, say a fixed point or a cycle. Therefore, for $\mu < \mu_c$, the behaviour of the system is very regular, at least for initial conditions not too far from A_1. When μ is increased slightly past μ_c, A_1 loses its stability and orbits now converge to a more complicated invariant set A_2, such that $A_1 \subset A_2$. However, for μ slightly larger than μ_c, the observed sequences generated by G_μ are quite similar to those prevailing when $\mu < \mu_c$, and their dynamics are mostly regular with occasional bursts of irregular behaviour. The frequency with which these bursts occur depends, on average, on the distance between μ and μ_c. Intermittency is then characterised by the following two features:

(i) a local/global discontinuous (or catastrophic) bifurcation at $\mu = \mu_c$, leading to the disappearance of a 'small' attractor A_1 (characterised

---

[3] Of course the reasoning that follows could be restated inverting the relevant inequality signs.

by regular dynamics) and the appearance of a new 'large' attractor A_2 characterised by complex dynamics and containing the locus of the former;

(ii) if we consider a typical orbit on A_2, the probability for the state of the system to be in the subset A_1 of A_2 changes *continuously* and inversely with $|\mu - \mu_c|$ and it tends to one as $|\mu - \mu_c| \to 0$.

There exist three basic types of generic bifurcations (schematically represented in figure 8.3) leading to the phenomenon of intermittency for maps, namely: the fold (or saddle-node) bifurcation; the subcritical Neimark–Sacker bifurcation; the subcritical flip bifurcation. These are sometimes called, respectively, type I, type II and type III intermittency. Notice that in all three types of intermittency we have a local bifurcation of discontinuous (catastrophic) kind whose effects, however, are both local (changes in the properties of a fixed point) and global (changes in the orbit structure far from the fixed point). Thus are such cases referred to as local/global bifurcations.

remark 8.2 Intermittency is observed both in systems of differential and difference equations. In the former case, however, the phenomenon can be fully characterised by considering bifurcations of the Poincaré map of the flows. When studying a Poincaré map of a flow it is useful to remember the following facts:

(1) A fixed point of the map corresponds to a limit cycle of the flow.
(2) An invariant closed curve of the map corresponds to an invariant torus of the flow.
(3) When a fold bifurcation occurs for the map, two fixed points of the map (one stable and one unstable) coalesce and disappear, and this corresponds to coalescence and disappearance of a stable and an unstable periodic orbit of the flow.
(4) A Neimark–Sacker bifurcation of the map corresponds to a bifurcation of a periodic orbit of the flow sometimes called 'secondary Hopf bifurcation'. If this bifurcation is supercritical, when the controlling parameter is, say, increased through the critical point, a limit cycle loses its stability and an attracting torus is born. Vice versa, if the bifurcation is subcritical, when we decrease the parameter through the critical point, an unstable limit cycle becomes stable and a repelling torus is born.

We shall not discuss the various types of intermittencies and their rather subtle technical details but concentrate on the simplest type I intermittency

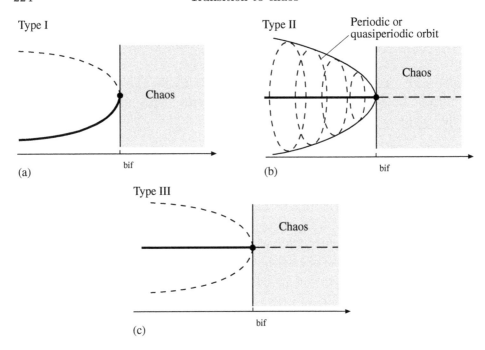

Fig. 8.3 Intermittency transition to chaos. Reproduced by permission of Cambridge University Press from Ott, *Chaos in Dynamical Systems* (1993)

that can be studied in relation to the familiar, one-dimensional logistic map $G(x_n) = \mu x_n(1 - x_n)$.

Consider the situation for values of the parameter μ in the vicinity of $\mu_c = 1 + 2\sqrt{2}$, that is, near the left boundary of the period-3 window (see figure 8.4(a)) . For $\mu < \mu_c$ the map G has two unstable fixed points, namely 0 and $(1 - 1/\mu)$. As μ *increases* past the critical value μ_c, the map acquires a stable and an unstable period-3 cycle. The three points of each cycle are stable (respectively, unstable) fixed points of the map G^3. If we reverse the procedure and *decrease* μ, at $\mu = \mu_c$, three pairs of (stable and unstable) fixed points of G^3 coalesce and disappear. The bifurcation occurring at μ_c is the familiar fold (saddle-node or tangent) bifurcation and is associated with the sudden, catastrophic disappearance, or appearance, of equilibria.

For $\mu > \mu_c$, the asymptotic dynamics generated by G for most initial points is a period-3 orbit, whereas, for values of μ slightly smaller than μ_c, we have a regular, almost periodic motion, occasionally interrupted by bouts of apparently chaotic behaviour. Although the overall motion is aperiodic, most iterates of the map are concentrated in the neighbourhood of the three points of the (disappeared) period-3 orbit, and the average duration of regular dynamics is a continuous (inverse) function of the distance $|\mu - \mu_c|$.

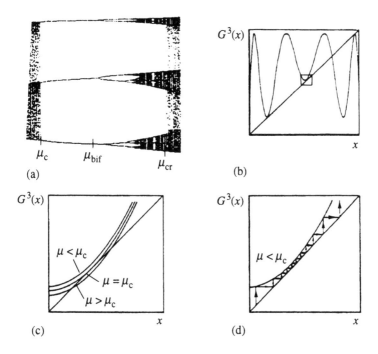

Fig. 8.4 Type I intermittency. Reproduced by permission of Cambridge
University Press

Different aspects of this behaviour are provided in figure 8.4(b), (c) and (d).
In particular, figure 8.4(d) suggests why there is a trace of the period-3 cycle
as μ is decreased slightly past μ_c. The motion of the system slows down
in the narrow channel in the vicinity of the locus where a stable periodic
point of G (a stable fixed point of G^3) existed. After a number of apparently
regular iterations (the greater the number, the nearer μ is to μ_c), the system
leaves the neighbourhood and wanders away in an irregular fashion until it
is reinjected into the channel between the curve of the map and the bisector,
and so on.

Another interesting aspect of the qualitative change in the orbit structure is visible in figure 8.4(a). If μ is increased further past μ_c, at a certain
value μ_{bif}, there begins a series of period-doubling bifurcations leading to
orbits of period $3 \cdot 2^i$ $(i = 1, 2, 3, \ldots)$, and eventually to the so-called **periodic chaos** (or noisy periodicity), i.e., chaotic behaviour restricted to
three narrow bands, each of them visited periodically.[4] Past another critical value μ_{cr}, the situation changes, again discontinuously, and the three

---

[4]Notice that in this case the map G is not topologically mixing on the chaotic set (cf. remark
6.4). Accordingly, the iterated map G^3 is not topologically transitive on that set.

bands suddenly merge and broaden to form a single chaotic region. This occurrence is sometimes called *crisis*, an interesting phenomenon to which we devote the next section.

8.3 Crises

Broadly speaking, crises occur as a consequence of a collision between an attractor and a saddle-like object such as a saddle fixed or periodic point, or their stable manifolds. There exist three types of crises involving chaotic attractors which occur when a control parameter passes through a critical value, as follows:

(i) The sudden increase in size of a chaotic attractor. This occurs when the collision takes place in the interior of the basin of attraction and is sometimes called **interior crisis**. Numerical evidence of interior crises has been found for a variety of dynamical systems in continuous and discrete time. A detailed discussion of the mechanism leading to this type of crisis is provided by Alligood *et al.* (1997), pp. 415–6. See also Ott (1993), pp. 277–84.

(ii) The sudden merging of two (or more) chaotic attractors or two (or more) separate pieces of a chaotic attractor, taking place when they simultaneously collide with an unstable fixed or periodic point (or its stable manifold). Crises of type (ii) have been numerically observed, for example, in the Hénon map. This is the type of crisis occurring for the logistic map when the controlling parameter μ reaches the value μ_{cr} (see figure 8.4(a)).

(iii) The sudden annihilation of a chaotic attractor. This happens when the collision occurs on the boundary of the basin of attraction (for short, on the **basin boundary**[5]). This crisis is sometimes called a **boundary crisis**. The colourful term 'blue sky catastrophe' was also used in Abraham and Marsden (1978), to emphasise the fact that when the control parameter is varied, an attractor of the system suddenly disappears into the blue, and the orbits originating in what used to be the basin of attraction of the destroyed attractor wander away, and possibly move towards another remote attractor. Boundary crises have been observed numerically, for example, in the Ikeda and Chua models (see Alligood *et al.*, 1997, pp. 202, 375).

---

[5] Let bd(A) indicate the **boundary** of a set A and $N(\epsilon, x)$ is the set of points whose distance from x is less than ϵ. Then we define

$$\text{bd}(A) = \{x| \text{ for every } \epsilon > 0, N(\epsilon, x) \text{ contains points both in } and \text{ not in } A\}.$$

Notice that a point $x \in \text{bd}(A)$ may or may not be in A.

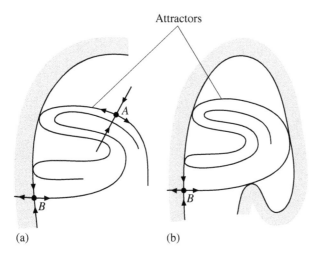

Fig. 8.5 Tangency crises at critical parameter values: (a) heteroclinic; (b) homoclinic. Reproduced by permission of Cambridge University Press from Ott, *Chaos in Dynamical Systems* (1993)

(Naturally, we could reverse the processes (i)–(iii), by reversing also the direction of change of the controlling parameter, and have, respectively, decrease, splitting, creation of chaotic attractors.)

Some particularly interesting boundary crises involve heteroclinic or homoclinic collisions. We have a **heteroclinic tangency crisis** when there occurs a collision between the stable manifold of a fixed or periodic point A of the saddle type, and the unstable manifold of another saddle B, as in figure 8.5(a). On the other hand, we have a **homoclinic tangency crisis** when there occurs a collision between the stable and the unstable manifolds of the same saddle B, as in figure 8.5(b).

Even the simplest homoclinic crises are very complex events which we cannot discuss here in detail. To give an idea of the complexity, consider the 'stroboscopic' depiction of a possible sequence of steps leading to creation (or annihilation) of a chaotic attractor through a homoclinic bifurcation in figure 8.6.[6] The figure represents five snapshots of a two-dimensional Poincaré map of a three-dimensional flow, each of them corresponding to a certain value of a control parameter. Reading from left to right we can identify (intuitively, not rigorously) the following steps:

---

[6]Notice that the illustrations are necessarily incomplete and collections of points have been replaced by continuous curves for expediency.

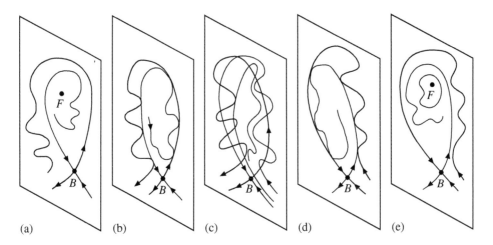

Fig. 8.6 A blue sky catastrophe for a three-dimensional flow (Poincaré map).
Reproduced by permission of Aerial Press from Abraham and Shaw, *Dynamics –
the Geometry of Behavior. Part 4: Bifurcation Behavior* (1988)

(a) There are two fixed points of the Poincaré map, a saddle B and an
unstable focus F (corresponding to two unstable limit cycles of the
flow).

(b) As the control parameter increases, the stable and unstable manifolds
of the saddle approach one another. When they touch, generically they
must have infinitely many points of tangency, corresponding to infinitely
many folds of the respective invariant sets of the three-dimensional flow.
The contact is one-sided, the unstable manifold of B encircling its stable
manifold. Transients here are very complex owing to the presence of
horseshoe-like structures.

(c) For a still higher value of the parameter, a chaotic *attractor* appears,
whose Poincaré map is a fractal collection of points lying on a curve with
a characteristic shape, called a 'bagel' by Abraham and Shaw (1988).

(d) Here the situation is symmetrical to that in *(b)*, with the stable manifold of B now encircling the unstable one. The range of values of the
parameter corresponding to Poincaré maps *(b)–(d)* is called the **tangle
interval**.

(e) When the parameter value is further increased out of the tangle interval, the tangencies of the invariant sets disappear and orbits, starting
inside the region encircled by the stable manifold of the saddle B, spiral
inwards, towards a stable torus.

We would like to mention two important additional features of interior
crises, on the one hand, and boundary crises, on the other. Recall that

the crises take place at a critical value μ_c of a controlling parameter μ and that we assume the attractor is enlarged or created when μ increases through μ_c and is reduced or annihilated when μ decreases through μ_c. In an interior crisis, for μ near μ_c ($\mu < \mu_c$), we typically have the phenomenon of intermittency as discussed above, and the average time distance between bursts of irregular (chaotic) dynamics will be the larger, the closer μ is to μ_c. On the contrary, following a boundary crisis, for μ slightly smaller than μ_c, we have chaotic transients, which will be the longer the closer μ is to μ_c. If transients are sufficiently long, we may speak of **transient chaos**. However, after the transients have died out, the system will move away permanently, possibly converging to some remote attractor. In numerical simulations, if transient chaos lasts a very long time, it may be confused with a chaotic attractor.

Secondly, and very importantly, boundary crises may be related to the phenomenon of hysteresis. Broadly speaking, we say that **hysteresis** occurs when the effects on the dynamics of a system due to changes (in one direction) in a certain parameter value are *not* reversed when the parameter is changed back to the original value. A simple example (not involving chaos) is sketched in figure 8.7 where equilibrium values of the state variable x are on the ordinate, parameter values on the abscissa. We assume that for all positive initial conditions:

(i) for $0 < \mu < \mu_c$, the system is characterised by a unique, stable equilibrium $\bar{x}_1 = 0$;

(ii) for $\mu > \mu_c$ there are three equilibria, $\bar{x}_1 = 0$ which is still stable, $\bar{x}_2$ stable, and $\bar{x}_3$ unstable;

(iii) at $\mu = \mu_c$, the two nontrivial equilibria coalesce and a saddle-node bifurcation occurs.

Suppose now that we fix $\mu > \mu_c$. For initial conditions inside or above the parabolic curve of equilibria ($x_0 > \bar{x}_3$), the system converges to $\bar{x}_2$. As μ is decreased past μ_c, the system, starting from the same initial conditions as before, converges to zero and remains arbitrarily close to it, for any value of μ, $0 < \mu < \mu_c$. If the change in μ is now reversed and the parameter is increased back past μ_c, the system will continue to converge to zero for sufficiently small initial conditions. To get the system back to equilibrium $\bar{x}_2$, increasing the parameter is not enough, the system must be given a push to move it into the basin of attraction of $\bar{x}_2$.

We can explain hysteresis in the context of crises as follows. Suppose that in a boundary crisis, when the controlling parameter is decreased past the critical value, a chaotic attractor A disappears and, after more or less long transients, orbits converge to a remote attractor B, which may or may not

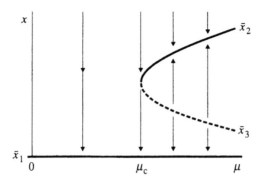

Fig. 8.7 A simple example of hysteresis

be chaotic. Generically, the attractors A and B will be disjoint and the two sets will not cease to be attracting simultaneously for the same value of μ. Consequently, if we increase μ past μ_c, we may have a situation in which, although the set A is again attracting, the observed orbits still converge to B, unless an exogenous shock pushes the system back into the basin of attraction of A.

remark 8.3 Some of the phenomena discussed in this section, and in particular the sudden creation or annihilation of attracting sets when a parameter is changed and saddle collisions occur, do not necessarily involve *chaotic* attractors, but could be similarly argued in relation to simpler periodic or quasiperiodic attracting sets. A simple example of a boundary crisis involving a limit cycle of a two-dimensional flow is depicted in figure 8.8 which is derived from the study of the well-known van der Pol system of two ordinary, scalar differential equations,[7] namely:

$$\dot{x} = ky + \mu x(b - y^2)$$
$$\dot{y} = -x + \mu.$$

For values of μ smaller than a critical value μ_c, there exists a saddle point and a stable limit cycle. As we increase μ, the limit cycle moves progressively nearer the saddle point until, at $\mu = \mu_c$, one branch of the unstable set and one branch of the stable set of the saddle merge with the limit cycle, that is, a homoclinic connection occurs while the cycle disappears. (The homoclinic connection itself can be thought of as an infinite-period cycle.)

Past μ_c, all the orbits originating in a neighbourhood of the saddle point, below as well as above its stable branches, wander away possibly moving

[7]The description that follows is based on Thompson and Stewart (1986), pp. 268–72.

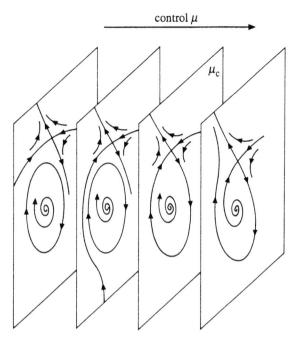

control μ

μ_c

Fig. 8.8 Boundary crisis involving a cycle. Reproduced by permission of John
Wiley & Sons from Thompson and Stewart, *Nonlinear Dynamics and Chaos*
(1986)

toward another attractor. Notice that, unlike the case of intermittency, this
attractor does not contain the locus of the 'old' one, and may be far from
it. However, the 'new' attractor 'inherits' the basin of attraction of the 'old'
one (which has now disappeared).

As we mentioned before the boundary crisis (or blue sky catastrophe) can
be accompanied by hysteresis. A schematic representation of a hypothetical
case is illustrated in the diagram of figure 8.9, where on the abscissa we have
the controlling parameter μ and on the ordinate (a projection of) the state
space. For $\mu > \mu_2$, the only attractor is a stable limit cycle. If we reduce μ
continuously, at μ_1 the limit cycle touches the saddle branch, a crisis occurs,
the limit cycle is destroyed and the system moves up towards the remote
stable node. If we reverse the change, however, and we *increase* μ past μ_1,
although a limit cycle is created, the node is not destroyed and continues
to attract points in its basin of attraction until we reach μ_2, at which point
a fold catastrophe takes place, the node disappears and the system jumps
back to the limit cycle. Thus, sequences of changes in the global dynamics
of the system are different according to whether the control parameter is
increased or decreased, which is the essence of hysteresis.

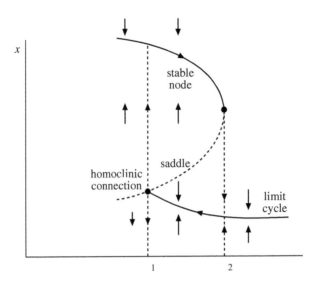

Fig. 8.9 Homoclinic connection and hysteresis. Reproduced by permission of John Wiley & Sons from Thompson and Stewart, *Nonlinear Dynamics and Chaos* (1986)

8.4 Quasiperiodicity and chaos

The idea that quasiperiodicity is the fundamental intermediate step in the route to chaos, is a long-standing one. As early as 1944, the Russian physicist Landau suggested that turbulence in time, or chaos, is a state approached by a dynamical system through an infinite sequence of Hopf bifurcations which take place when a certain parameter is changed. Thus, according to this scenario, the dynamics of the system would be periodic after the first bifurcation, and quasiperiodic after the successive bifurcations, with an ever-increasing order of quasiperiodicity (a higher number of oscillators with incommensurable frequencies), in the limit leading to turbulent, chaotic behaviour (see figure 8.10).

The conjecture that Landau's is the only, or even the most likely route to chaos was questioned on the basis of the following two basic results:

(i) From a theoretical point of view, Ruelle and Takens (1971) and Newhouse *et al.* (1978) proved that systems with 4-torus (or even 3-torus) attractors can be converted to systems with chaotic attractors by means of arbitrarily small perturbations.[8]

[8]The results given by Newhouse, Ruelle and Takens are sometimes said to define a possible route from quasiperiodicity to chaos. However, those authors showed only that arbitrarily near (in the appropriate space) to a vector field having a three- (or four-) frequency quasiperiodic attractor, there exist vector fields having chaotic attractors. They did not prove that there exists an

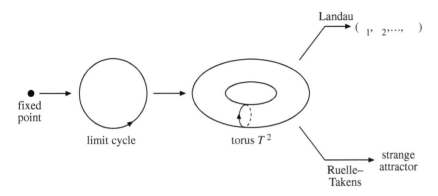

Fig. 8.10 Landau and Ruelle–Takens routes to chaos

(ii) Lorenz's (1963) work showed that complexity (in the sense of a large dimension of the system) is not a necessary condition for chaos to occur, and that low-dimensional systems are perfectly capable of producing chaotic output.

Although experimental and numerical results suggest the possibility of a direct transition from quasiperiodicity to chaos, mathematically there remain some open questions which shall not be discussed here. However, some aspects of the 'quasiperiodic route to chaos' are related to the Neimark–Sacker bifurcation of discrete-time dynamical systems discussed in chapter 5, and we devote the next few paragraphs to them.

In section 5.4, we considered Neimark–Sacker bifurcations for a *one-parameter* family of maps such that for a certain critical value of the controlling parameter, the modulus of a pair of complex conjugate eigenvalues of the Jacobian matrix of the map in $\mathbb{R}^2$ goes through the unit circle. Assuming away the case of *strong resonance*, we saw that the dynamics of the system near the bifurcation point can be studied by means of a fixed rotation map of the circle. To carry the discussion a little further we reformulate the problem as a *two-parameter* family of maps by means of the following so-called **sine circle map**, first introduced by Arnold (1965)

$$\theta_{n+1} = G(\theta_n) = \theta_n + \omega + \frac{k}{2\pi}\sin(2\pi\theta_n) \quad \text{mod } 1 \qquad (8.1)$$

where θ is an angular variable and G is a map of the (parametrised) circle, in the additive notation and $0 \le \omega \le 1$.

arc in control space, along which a transition from quasiperiodicity to chaos can be realised. For more details see Bergé *et al.* (1984), pp. 159–91; Thompson and Stewart (1986), pp. 287–8; Schuster (1989), pp. 145–85; Ott (1993), pp. 202–3 and the bibliography quoted therein.

remark 8.4 Maps like (8.1) can be interpreted as models for systems characterised by a pair of oscillators with two different frequencies, coupled non-
linearly. A prototypical example of such systems is the periodically forced
oscillator investigated by van der Pol which provided one of the earliest
examples of chaotic dynamics. In this case, the parameters ω and k can
be interpreted as the ratio between natural and forcing frequencies and the
forcing amplitude, respectively.

For $k = 0$, (8.1) reduces to the fixed rotation of the circle discussed in
chapter 5 (5.30), and in this case ω is equal to the rotation number ρ. For
$0 \leq k \leq 1$ the dynamics of the system are periodic or quasiperiodic according to whether ρ is rational or irrational and, in general, ρ will be a function
of both ω and k. Let us now define the 'periodic set' and 'quasiperiodic set'
as, respectively,

$$P = \{\omega(k) \in (0,1) \mid \rho \text{ is rational}\}$$
$$Q = \{\omega(k) \in (0,1) \mid \rho \text{ is irrational}\}$$

with $k \in [0,1]$. If m denotes the Lebesgue measure on $[0,1]$, for $k = 0$
we have $m(P) = 0$ and $m(Q) = 1$. If $k > 0$ but small, Q is a Cantor
set (including no intervals) and $m(Q) > 0$. However, arbitrarily close to
each $\omega \in Q$ there exist *intervals* of $\omega \in P$ (which implies $m(P) > 0$) over
which ρ remains locked to some fixed rational value p/q and therefore, over
each interval, ρ is independent of ω. This phenomenon is known as **mode-**,
phase- or **frequency-locking**. The regions of frequency-locking in the
(k, ω) plane have characteristic shapes, known as 'Arnold tongues', appearing as black regions in figure 8.11. At $k = 1$, $m(P) = 1$ and $m(Q) = 0$
and we can say, broadly speaking, that for $0 \leq k \leq 1$, quasiperiodicity
prevails for small k and periodicity for large k.

For $k > 1$ the map G ceases to be invertible, the circle is folded onto
itself, Arnold tongues can overlap, and chaotic behaviour becomes possible.
However, for the reasons we have just explained, the crossing of the $k = 1$
line (sometimes referred to as 'transition to chaos') will typically take place
at a value of ω such that $\rho(1, \omega)$ is rational and the dynamics are periodic.
Therefore, in moving from periodicity to chaos, the system would typically
follow one of the canonical (one-parameter) routes discussed before. However if *both parameters ω and k are controlled simultaneously*, the line $k = 1$
can be crossed at a value of ω belonging to (the zero measure) quasiperiodic
set Q. By carefully controlling ω and k, therefore, a path can be followed
in the (ω, k) parameter space that avoids the frequency-locking regions and
moves directly from quasiperiodicity to a chaos. The reader can study the

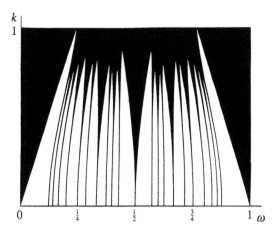

Fig. 8.11 Frequency-locking. Reproduced by permission of Cambridge University Press from Ott, *Chaos in Dynamical Systems* (1993)

rather complicated details of this route in Mori and Kuramoto (1998), pp. 174–82.

Exercises

8.1   Check that the Schwarzian derivative of the maps $G(x_n) = \mu x_n(1 - x_n)$ and $G(x_n) = \mu - x_n^2$ is negative.

8.2   Consider again the logistic map. Find the ω-limit set, Λ, to which most points in $[0, 1]$ converge, given the following parameter values: $\mu = 0.8, 2, 3.2, 4$. For each of these values of the parameter, find some initial values that do not converge to the respective Λ and give their limit sets.

8.3   Let G be a continuous map on $\mathbb{R}$.

(a) Suppose G has a prime period-32 cycle. What other cycles exist for G?

(b) Does the existence of a prime period-1024 cycle ensure the existence of a period-24 cycle?

(c) If a prime period-44 cycle is found for G, should it be expected to find (a) a period-40 cycle, (b) a period-32 cycle, (c) a period-12 cycle?

8.4   Consider the period-doubling scenario for the logistic map. Approximate the value of the parameter μ_3 corresponding to the third flip bifurcation given $\mu_1 = 3$, $\mu_2 = 1 + \sqrt{6}$. How does that value compare to the actual value $\mu_3 \approx 3.544\,090$ (approximated to the sixth decimal place)?

8.5   In exercise 5.8(a) it was demonstrated that the map $G(x_n) = \mu - x_n^2$ undergoes a first flip bifurcation at $(\mu_1, \bar{x}) = (3/4, 1/2)$:

   (a) Calculate the parameter value of μ_2 at which the period-2 cycle loses its stability through a second flip bifurcation by considering that the eigenvalue of G^2, calculated at one of the periodic points of the cycle must equal -1.
   (b) Approximate μ_3. In this case the actual value is $\mu_3 = 1.368\,099$ (to the sixth decimal place).
   (c) Approximate the Feigenbaum constant δ using these values: $\mu_6 = 1.400\,829$ (period-64 cycle); $\mu_7 = 1.401\,085$ (period-128 cycle); $\mu_8 = 1.401\,140$ (period-256 cycle).

9

The ergodic approach

In previous chapters we discussed dynamical systems mainly from a geometrical or topological point of view. The geometrical approach is intuitively appealing and lends itself to suggestive graphical representations. Therefore, it has been tremendously successful in the study of low-dimensional systems: continuous-time systems with one, two or three variables; discrete-time systems with one or two variables. For higher-dimensional systems, however, the approach has encountered rather formidable obstacles and rigorous results and classifications are few. Thus, it is sometimes convenient to change perspective and adopt a different approach, based on the concept of measure, and aimed at the investigation of the statistical properties of ensembles of orbits. This requires the use and understanding of some basic notions and results, to which we devote this chapter. The ergodic theory of dynamical systems often parallels its geometric counterpart and many concepts discussed in chapters 3–8, such as invariant, indecomposable and attracting sets, attractors, and Lyapunov characteristic exponents will be reconsidered in a different light, thereby enhancing our understanding of them. We shall see that the ergodic approach is very powerful and effective for dealing with basic issues such as chaotic behaviour and predictability, and investigating the relationship between deterministic and stochastic systems.

From the point of view of ergodic theory, there is no essential difference between discrete- and continuous-time dynamical systems. Therefore, in what follows, we develop the discussion mostly in terms of maps, mentioning from time to time special problems occurring for flows.

9.1 Ergodic measures

9.1.1 Some elementary measure theory

After these preliminary considerations, we now need to define a certain number of concepts and methods which will be used in the sequel of this chapter. Ergodic theory discusses dynamical systems in terms of two fundamental mathematical objects: a **measure space** $(X, \mathcal{F}, \mu)$ and a **measure-preserving** map $G : X \to X$ (or, for short, $X \hookleftarrow$). In ergodic theory, a measure-preserving map G acting on a measure space is often called an **endomorphism**. If, in addition, G is invertible, it is called an **automorphism**. In this context, G is also called **transformation**.

X is some arbitrary space. $\mathcal{F}$ is a collection of subsets of X such that:

(i) $X \in \mathcal{F}$;
(ii) if $A \in \mathcal{F}$, then the complement of A, $X \setminus A \in \mathcal{F}$;
(iii) given a finite or infinite sequence $\{A_i\}$ of subsets of X such that $A_i \in \mathcal{F}$, their union $\cup_i A_i \in \mathcal{F}$.

From properties (i) and (ii), it follows that

(iv) the empty set $\emptyset \in \mathcal{F}$; and
(v) $\cap_i A_i \in \mathcal{F}$, i.e., $\mathcal{F}$ is closed under the operation of intersection.

The set $\mathcal{F}$ endowed with these properties is called a **σ-algebra**. If X is a metric space, the most natural choice for $\mathcal{F}$ is the so-called *Borel σ-algebra*, $\mathcal{B}$. By definition, this is the smallest σ-algebra containing open subsets of X, where 'smallest' means that any other σ-algebra that contains open subsets of X also contains any set contained in $\mathcal{B}$. In what follows, unless stated otherwise, it will always be assumed that $\mathcal{F} = \mathcal{B}$.

The third element of the triplet $(X, \mathcal{F}, \mu)$ is a **measure**. Broadly speaking, a measure belongs to the class of **set functions**, i.e., functions assigning numerical values to sets. In order for μ to be a measure, values must be assigned in a reasonable manner, in the sense that μ must satisfy the following requirements:

(i) $\mu(A) \geq 0$ for all $A \in \mathcal{F}$, that is, μ must assign nonnegative values (possibly $+\infty$);
(ii) $\mu(\emptyset) = 0$, that is, the empty set must have zero measure;
(iii) if $\{A_i\}$ is a sequence of pairwise disjoint sets, then

$$\mu \left(\bigcup_{i=\mathbb{N}} A_i \right) = \sum_{i=1}^{\infty} \mu(A_i).$$

That is to say, if a set is the union of a countable number of disjoint pieces, the sum of the values assigned to its subsets should be equal to the value assigned to the set as a whole.

For a set $A \in \mathcal{F}$, the quantity $\mu(A)$ is called the measure of A and the sets $A \in \mathcal{F}$ are called **measurable sets** because, for them, the measure is defined. The integral notation[1]

$$\mu(A) = \int_A \mu(dx) \quad A \in \mathcal{B}(X)$$

is often used, where the expression on the RHS is the Lebesgue integral with respect to the measure μ.

Here we are interested in finite measure spaces, such that $\mu(X) < \infty$. In particular, if $\mu(X) = 1$ the measure space is said to be 'normalised' or 'probabilistic' and $0 \leq \mu \leq 1$ is called a **probability measure**. The smallest closed subset of X that is μ-measurable and has a μ-null complement is called the **support of** μ and denoted by $\mathrm{supp}(\mu)$ and we have

$$[x \in \mathrm{supp}\ (\mu)] \iff [\mu(B_r(x)) > 0 \quad \forall r > 0]$$

where $B_r(x)$ denotes a ball of radius r centred on x.

A transformation G is said to be **measurable** if $[A \in \mathcal{F}] \Rightarrow [G^{-1}A = \{x \mid G(x) \in A\} \in \mathcal{F}]$. G is said to be **measure-preserving** with respect to μ or, equivalently, μ is said to be **G-invariant**, whenever $\mu(G^{-1}(A)) = \mu(A)$ for all sets $A \in \mathcal{F}$. Thus, G-invariant measures are compatible with G in the sense that sets of a certain size (in terms of the selected measure) will be mapped by G into sets of the same size.

remark 9.1 When we are dealing with flows, we say that a family $\phi_t(x)$ of measurable maps $X \hookleftarrow$ preserves the measure μ if

$$\mu(A) = \mu\left[\phi_t^{-1}(A)\right]$$

for all t and all measurable subsets $A \subset X$.

In the applications with which we are concerned, X denotes the state space and usually we have $X \subset \mathbb{R}^m$; the sets $A \in \mathcal{F}$ denote configurations of the state space of special interest, such as fixed or periodic points, limit cycles, strange attractors, or subsets of them. The transformation G is the law governing the time evolution of the system. We shall often refer to

---

[1]Notice that different, equivalent notations for the Lebesgue integral are in use, thus if f is an integrable function on X, we can write equivalently

$$\int f d\mu = \int_X f(x) d\mu(x) = \int_X f(x)\mu(dx)$$

where $x \in X$ and the second and third forms are used to emphasise the integrand's argument.

the quadruplet $(X, \mathcal{F}, \mu, G)$, or even to the triplets $(X, \mathcal{F}, \mu)$ or $(X, \mathcal{F}, G)$ as 'dynamical systems'. When the σ-algebra $\mathcal{F}$ has been specified but the measure μ has not, then $(X, \mathcal{F})$ is called a **measurable space**.

We concentrate the discussion on the study of invariant measures for reasons that can be explained as follows. As we want to study the statistical, or probabilistic properties of orbits of dynamical systems, we need to calculate averages over time. As a matter of fact, certain basic quantities such as Lyapunov characteristic exponents (which, as we noted in chapter 7, measure the rate of divergence of nearby orbits), or metric entropy (which, as we shall see below, measures the rate of information generated by observations of a system) can be looked at as time averages. For this purpose, it is necessary that orbits $\{x, G(x), G^2(x), \ldots\}$ generated by a transformation G possess statistical regularity and certain limits exist.

More specifically, we often wish to give a meaningful answer to the following question: how often does an orbit originating in a given point of the state space visit a given region of it?

To formulate the problem more precisely, consider a transformation G of the space X, preserving a measure μ, and let A be an element of $\mathcal{F}$. Then we define

$$V_n(x) = \# \left\{ i | 0 \leq i < n, G^i(x) \in A \right\} \tag{9.1}$$

and

$$v_n(x) = \frac{1}{n} V_n(x) \quad x \in X \tag{9.2}$$

where $\#\{\cdot\}$ means 'the number of elements in the set'. Thus, $V_n(x)$ denotes the number of 'visits' of a set A after n iterations of G and $v_n(x)$ the *average* number of visits.

Now let n become indefinitely large. We would like to know whether the limit

$$\hat{v}(x) = \lim_{n \to \infty} v_n(x) \tag{9.3}$$

exists. This can be established by means of a basic result in ergodic theory, known as the **Birkhoff–Khinchin (B-K) ergodic theorem**.[2] In its generality, the B-K theorem states that, if $G : X \hookleftarrow$ preserves a probability measure μ and f is any function μ-integrable on X, then the limit

$$\lim_{n \to \infty} \frac{1}{n} \sum_{i=0}^{n-1} f[G^i(x)] = \hat{f}(x) \tag{9.4}$$

---

[2]Proofs of the B-K ergodic theorem can be found in many texts on ergodic theory. See, for example, Katok and Hasselblatt (1995), pp. 136–7; Keller (1998), pp. 24–30.

exists for μ-almost every point[3] $x \in X$, $\hat{f}(x)$ is integrable and G-invariant, i.e., $\hat{f}(G(x)) = \hat{f}(x)$, wherever it is defined. Moreover, for each G-invariant set $A \in \mathcal{F}$, i.e., for A such that $G^{-1}(A) = A$

$$\int_A \hat{f} d\mu = \int_A f d\mu.$$

The reader can easily verify that, if we choose $f = \chi_A$, where χ_A denotes the so-called **characteristic function**, or **indicator** of A, namely:

$$\chi_A(x) = \begin{cases} 1, & \text{if } x \in A \\ 0, & \text{if } x \notin A, \end{cases}$$

then $(1/n) \sum_{i=0}^{n-1} f[G^i(x)] = v_n(x)$ and therefore the B-K theorem guarantees the existence of limit (9.3), μ-almost everywhere.

remark 9.2 The B-K theorem is a proposition about asymptotic behaviour of a dynamical system with respect to a *given* invariant measure μ. It says nothing about existence or uniqueness of such a measure. A result by Krylov and Bogolubov (see Keller, 1998, p. 15) establishes that any continuous map on a compact metric space must have at least one invariant measure. In general, we expect dynamical systems to have many invariant measures, especially if the system is characterised by complex behaviour. In such cases we are faced with the difficult problem of deciding which of them is relevant for the situation under investigation. This issue comes up again in section 9.3.

9.1.2 Invariant, ergodic measures

In general, however, the limit (9.4) in the B-K theorem depends on x, which means that the time average in question may be different for orbits originating from different initial states. This happens, for example, when the space X is decomposable under the action of G and there exist two subspaces X_1 and X_2, both invariant with respect to G, i.e., when G maps points of X_1 only to X_1 and points of X_2 only to X_2. In that case, clearly we expect different *histories* of the system, and consequently different time averages and statistical properties, depending on the initial conditions.

The dynamic decomposability of the system, which is a geometric or topological fact, is reflected in the existence of a G-invariant measure μ that is decomposable in the sense that it can be represented as a weighted average of invariant measures μ_1 and μ_2. That is, we have $\mu = \alpha\mu_1 + (1 - \alpha)\mu_2$,

---

[3]This means that the set of points of X for which the limit (9.4) does not exist is negligible, in the sense that the invariant measure μ assigns zero value to that set.

where $\alpha \in (0,1)$ and the measures μ_1 and μ_2, in turn, may or may not be decomposable.

We are not so much interested in the properties of a single orbit starting from a given initial point, but in the overall properties of ensembles of orbits originating from most initial conditions in certain regions of the space. Thus, it would be desirable that the average calculated along a certain, randomly chosen history of the system should be a typical representative of the averages evaluated over most of the other possible histories. There exists a fundamental class of invariant measures that satisfy this requirement, called **ergodic measures**.

remark 9.3 Notice that, although in the literature one commonly encounters the expression 'ergodic map', this makes sense only if it is clear with respect to what invariant measure the map is indeed ergodic.

Several equivalent characterisations of ergodicity exist, three of which we present below.

definition 9.1 *Given a dynamical system* $(X, \mathcal{F}, \mu, G)$, *the (normalised)* G-*invariant measure* μ *is said to be* **ergodic** *for* G *if:*

  (i) *whenever* $G^{-1}(A) = A$ *for some* $A \in \mathcal{F}$, *either* $\mu(A) = 1$ *or* $\mu(A) = 0$; *or equivalently,*
 (ii) *if* f *is an integrable function,* G-*invariant* μ-*almost everywhere, then* f *is constant* μ-*almost everywhere; or equivalently,*
(iii) *the function* $\hat{f}$ *defined in the B-K theorem is a constant, and we have*

$$\lim_{n \to \infty} \frac{1}{n} \sum_{i=0}^{n-1} f[G^i(x)] = \hat{f} = \int_X f(x)\mu(dx) \qquad (9.5)$$

for μ-*almost every* x.

The RHS of (9.5) is the expected value of the function f with respect to the measure μ and is sometimes denoted by $E\{f\}$. It can be interpreted as an average with respect to space, where the measure μ 'weighs' each element of the state space X.

Property (iii) of definition 9.1 states that, if μ is ergodic, then μ-almost everywhere the time average of any integrable function f is equal to its expected value. The situation is often described summarily by saying that 'the time average equals the space average'.

Again using $f = \chi_A$, where A is a measurable subset of X, the ergodicity of μ implies that the average number of visits to a region A of an infinite orbit originating from μ-almost every point is equal to the 'size' that the ergodic measure μ assigns to that region. In formal terms, we have

$$\lim_{n\to\infty} \frac{1}{n} \sum_{i=0}^{n-1} \chi_A(G^i(x)) = \int_X \chi_A(x)\mu(dx) = \int_A \mu(dx) = \mu(A) \qquad (9.6)$$

for μ-almost every $x \in X$.

As we shall see, however, the set of points for which (9.5) does *not* hold may well include *almost every point* with respect to another measure, in particular with respect to the Lebesgue measure.

remark 9.4 Ergodicity is a property of the pair (μ, G). Thus, we can (and do) use equivalently the phrase 'μ is ergodic for G' or 'G is ergodic for μ'.

Before proceeding further in our analysis, we need to discuss two special and extremely useful types of measure.

definition 9.2 *Let us fix a point $x \in X$. The **Dirac measure** centred on x is the probability measure μ that assigns value one to all the subsets A of X that contain x, and value zero to those subsets that do not contain it. Formally, we have*

$$\mu(A) = \begin{cases} 1 & \text{if } x \in A \\ 0 & \text{if } x \notin A \end{cases}$$

The Dirac measure (centred on x) is also called 'Dirac delta' and is usually denoted by δ_x.

The second type is known as a k-dimensional Lebesgue measure. Consider a set $A \subset \mathbb{R}^k$ defined by

$$A = \{x \equiv (x_1, x_2, \ldots, x_k) \in \mathbb{R}^k | a_i < x_i \le b_i \ a_i, b_i \in \mathbb{R}, \ i = 1, 2, \ldots, k\}$$

and called a **coordinate parallelepiped** in $\mathbb{R}^k$. Then the **k-dimensional Lebesgue measure of A**, denoted by $m^k(A)$, is given by

$$m^k(A) = \text{vol}^k(A)$$

where $\text{vol}^k(A)$ is the k-dimensional volume of A, given by

$$\text{vol}^k(A) = \prod_{i=1}^{k}(b_i - a_i).$$

The k-dimensional Lebesgue measure can be extended to arbitrary subsets of $\mathbb{R}^k$ by defining

$$m^k(A) = \inf\left\{\sum_{i=1}^{\infty} \text{vol}^k(A_i) \Big| A \subset \bigcup_{i=1}^{\infty} A_i\right\}$$

where $\{A_i\}$ is a collection of coordinate parallelepipeds covering A and the infimum is taken over all possible coverings (i.e. over all possible countable unions of coordinate parallelepipeds containing A).[4] It is natural to think of m^1 as the length of an interval of $\mathbb{R}$, m^2 as the area of a subset of $\mathbb{R}^2$, m^3 as the volume of a subset of $\mathbb{R}^3$ and so on and so forth.

Notice that if f is an integrable function on $X \subset \mathbb{R}^k$, then the Lebesgue integral with respect to m^k is often denoted by

$$\int_X f(x)dx$$

rather than by $\int_X f(x)m^k(dx)$. If f is also Riemann integrable the Lebesgue integral and the Riemann integral are the same.

remark 9.5 In the discussion of certain questions such as the properties of attractors, we use the notion of *subset of positive (or zero) Lebesgue measure of a d-dimensional manifold*. This can be made precise as follows. In the definition in chapter 3, appendix C, p. 98, we indicated that a manifold M is characterised by a system of local coordinate charts, which are homeomorphisms from open subsets $W_i \subset M$ to open subsets $U = \phi(W_i) \subset \mathbb{R}^d$. Then we can define the volume measure μ of an open subset $B \subset W_i$ as

$$\mu(B) = \int_{\phi(B)} \frac{dx}{V_\phi \left[\phi^{-1}(x)\right]}$$

(which does not depend on the choice of the coordinate chart ϕ). The function $V_\phi \left[\phi^{-1}(x)\right]$ is called the **unit volume function** (of a point $y = \phi^{-1}(x) \in M$), defined as

$$V_\phi(y) = |\det(V)|,$$

where V is a $(d \times d)$ matrix whose columns are the (normalised) tangent vectors at y. The heuristic interpretation of $V_\phi \left[\phi^{-1}(x)\right]$ is that it represents the ratio between the elementary volume dx in $\mathbb{R}^d$ and the corresponding volume in M. (For a more detailed discussion, cf. Lasota and Mackey, 1985, pp. 151–8.)

Some examples of invariant, ergodic measures will give the rather abstract notions discussed so far a more intuitive flavour and also establish a bridge between the geometric and the ergodic approach to dynamical systems.

EXAMPLE 1 Consider a dynamical system in $\mathbb{R}^m$, characterised by a unique, fixed point $\bar{x}$. In this case, the Dirac measure $\delta_{\bar{x}}$ is invariant and ergodic.

---

[4]Cf. the definition of cover in n. 1, p. 108.

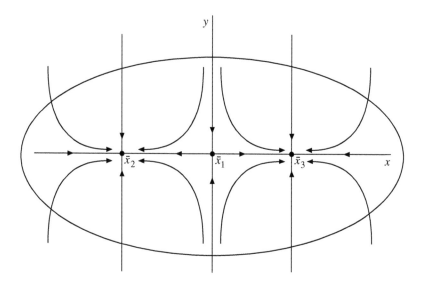

Fig. 9.1 A decomposable invariant set

EXAMPLE 2 Consider figure 9.1 representing the state space of the system of
ordinary difference equations $(\dot{x} = x - x^3; \dot{y} = -y)$ with fixed points $\bar{x}_1 = 0$,
$\bar{x}_2 = -1, \bar{x}_3 = 1$ and $\bar{y} = 0$. In this case, the measure $\mu = \alpha\delta_{\bar{x}_2} + (1-\alpha)\delta_{\bar{x}_3}$,
where $\alpha \in [0, 1]$, is invariant. However, as the reader can easily establish,
μ is not ergodic. It can be decomposed into the two measures $\delta_{\bar{x}_2}$ and $\delta_{\bar{x}_3}$
which are also invariant. Accordingly, the time averages of orbits starting
in the basin of attraction of $\bar{x}_2$ are different from those of orbits originating
in the basin of attraction of $\bar{x}_3$. The system is decomposable both from a
geometrical and from an ergodic point of view.

EXAMPLE 3 Consider a discrete-time dynamical system characterised by a
periodic orbit of period k, $\{x, G(x), \ldots, G^{k-1}(x)\}$. In this case, the measure
that assigns the value $1/k$ to each point of the orbit is invariant and ergodic.

EXAMPLE 4 Consider a continuous-time dynamical system characterised by
a limit cycle of period τ, $\Gamma = \{\phi_t(x) \mid 0 \le t < \tau\}$, where ϕ_t is the flow-map
associated to the solution of the system, τ is the period of the cycle and x is
any point on Γ. In this case, the following measure is invariant and ergodic:

$$\mu = \frac{1}{\tau} \int_0^\tau \delta_{\phi_t(x)} dt.$$

This means that the probability is spread evenly over the cycle Γ, which is
the support of μ, according to the time parameter.

EXAMPLE 5 Consider the map of the circle G_C (multiplicative notation) discussed in chapter 4 (see equation (4.11)), where $G_C : S^1 \to S^1$ and

$$z_{n+1} = G(z_n) = c z_n \quad c = e^{i2\pi\alpha} \; \alpha \in [0,1).$$

If we consider the map $\theta : \mathbb{R} \to S^1$, $\theta(x) = e^{i2\pi x}$, we can see that G_C preserves the measure $\hat{m} = m \circ \theta^{-1}$ where m is the Lebesgue measure on the interval. To prove this we must show that for any measurable subset $C \subset S^1$

$$m\left[\theta^{-1}(C)\right] = m\left(\theta^{-1}\left[G_C^{-1}(C)\right]\right).$$

But this is true because G_C^{-1} takes any arc of the circle and rotates it in a clockwise manner without changing its length and $m \circ \theta^{-1}$ maps subsets of S^1 to subsets of the interval and assigns to each of them a value equal to its length. Consequently, the measure $m \circ \theta^{-1}$ is preserved by G_C. $\hat{m}$ is known as the **Lebesgue measure on the circle**. If α is irrational, the measure $\hat{m} = m \circ \theta^{-1}$ is ergodic.

EXAMPLE 6 When dynamical systems are chaotic, invariant measures seldom can be defined in a precise manner. This difficulty parallels that of precisely locating chaotic invariant sets. We next discuss a well-known case in which the formula for the density of an invariant ergodic measure can be written exactly, namely the logistic map $G_4 : [0,1] \hookleftarrow$, $x_{n+1} = G_4(x_n) = 4x_n(1 - x_n)$.

In chapter 6, section 6.5 we showed that G_4 is topologically conjugate to the tent map G_Λ via the homeomorphism $\theta(y) = \sin^2(\pi y/2)$ and, in particular, that

$$\theta \circ G_\Lambda = G_4 \circ \theta. \tag{9.7}$$

Equation (9.7) implies that, if I is a subset of $[0,1]$, the sets

$$\theta^{-1} G_4^{-1}(I) = \{x \in [0,1] \mid G_4[\theta(x)] \in I\}$$
$$G_\Lambda^{-1}\theta^{-1}(I) = \{x \in [0,1] \mid \theta[G_\Lambda(x)] \in I\}$$

are the same, that is,

$$\theta^{-1}\left[G_4^{-1}(I)\right] = G_\Lambda^{-1}\left[\theta^{-1}(I)\right]. \tag{9.8}$$

Consider now that for any measurable subset $I \subset [0,1]$

$$G_\Lambda^{-1}(I) = I_1 \bigcup I_2$$

where I_1 and I_2 are two subsets whose lengths are equal to half the length of I. From this we conclude that G_Λ preserves the Lebesgue measure on $[0, 1]$.

We can also establish that G_4 preserves the measure $m \circ \theta^{-1}$. This requires that for any measurable subset I of $[0, 1]$, we have

$$m \left[\theta^{-1}(I) \right] = m \left(\theta^{-1}[G_4^{-1}(I)] \right) = m \left(G_\Lambda^{-1} \left[\theta^{-1}(I) \right] \right) \qquad (9.9)$$

the last equality deriving from (9.8). But (9.9) is true because G_Λ preserves the Lebesgue measure.

Then the still unspecified G_4-invariant measure μ must satisfy

$$\mu(I) = \int_{\theta^{-1}(I)} dy$$

for any measurable subset I of the interval. By making use of a standard change-of-variable rule (cf. Billingsley, 1965, p. 192) we have

$$\mu(I) = \int_{\theta^{-1}(I)} dy = \int_I |(\theta^{-1})'(x)|\, dx.$$

Considering that

$$(\theta^{-1})'(x) = \frac{1}{\theta'[y(x)]}$$

we have

$$\mu(dx) = \frac{dx}{\pi\sqrt{x(1-x)}}. \qquad (9.10)$$

9.2 Lyapunov characteristic exponents revisited

We can use the ideas developed in the previous section to reformulate the definition of Lyapunov characteristic exponents and reconsider the question of the existence of the limits defining LCEs and their dependence on initial conditions, which was left open in chapter 7. We start with the simpler case of a one-dimensional map G preserving a measure μ.

Recalling our discussion of the Birkhoff–Khinchin theorem and ergodicity of measures, and putting $f(x) = \ln |G'(x)|$, we can conclude that the limit defining the LCE of G evaluated along an orbit starting at x_0, that is,

$$\lambda(x_0) = \lim_{n \to \infty} \frac{1}{n} \sum_{i=0}^{n-1} \ln \left| G'[G^i(x_0)] \right|$$

exists μ-almost everywhere. Moreover, if μ is ergodic, we have

$$\lambda = \lim_{n \to \infty} \frac{1}{n} \sum_{i=0}^{n-1} \ln \left| G'[G^i(x_0)] \right| = \int \ln |G'(x)| \mu(dx) \qquad (9.11)$$

where the integral is taken over the state space. Notice that the fact that $f(x)$ may be undefined for some points (for the logistic map at $x = 1/2$, for example) is not a problem here, as long as $f(x)$ is μ-integrable. Equation (9.11) can be given a heuristic interpretation by saying that the LCE is the local stretching determined by the (logarithm of the) slope of the curve of the map G, weighted by the probability of that amount of stretching according to the measure μ. Sets of point of zero measure 'do not count'. All this can be illustrated by two simple examples.

EXAMPLE 7 The asymmetric tent map

$$G_{\hat{\Lambda}} : [0, 1] \to [0, 1]$$

$$G_{\hat{\Lambda}}(x) = \begin{cases} x/a & \text{for } 0 \le x \le a \\ (1 - x)/(1 - a) & \text{for } a < x \le 1. \end{cases} \qquad (9.12)$$

It is easily seen that (9.12) preserves the length of subintervals of $[0, 1]$ and therefore it preserves the Lebesgue measure. Then, recalling our comment in section 9.1.1 and putting $m(dx) = dx$, we have

$$\lambda = \int_0^1 \ln |G'_{\hat{\Lambda}}(x)| dx = \int_0^a \ln \left(\frac{1}{a} \right) dx + \int_a^1 \ln \left(\frac{1}{1-a} \right) dx =$$

$$= a \ln \left(\frac{1}{a} \right) + (1 - a) \ln \left(\frac{1}{1-a} \right). \qquad (9.13)$$

If $a = 1/2$, we are in the case of the symmetric tent map and $\lambda = \ln 2$.

EXAMPLE 8 As usual G_4 denotes the logistic map at the parameter value 4, $G_4(x) = 4x(1 - x)$. Recalling example 6, we have for almost all points x,

$$\lambda = \int_{[0,1]} \frac{\ln |G'_4(x)|}{\pi[(x(1-x)]^{1/2}} dx = \int_{[0,1]} \frac{\ln |4 - 8x|}{\pi[x(1-x)]^{1/2}} dx = \ln 2. \qquad (9.14)$$

However, notice that, if we choose the special initial point $x_0 = 0$, we have $\lambda(x_0) = \lim_{n \to \infty} \frac{1}{n} \sum_{i=0}^{n-1} \ln |G'_4[G^i(x_0)]| = \ln |G'_4(x_0)| = \ln 4$.

In the general m-dimensional case, the B-K theorem cannot be applied to the analysis of LCEs as simply and directly as in the one-dimensional case. Broadly speaking, the root of the difficulty is the fact, when G is multi-dimensional, the quantities $DG(x)$ (where D, as usual, is the partial derivative operator) are not scalars but noncommuting matrices and consequently their product along an orbit

$$DG^n(x_0) = \prod_{i=0}^{n-1} DG(x_i)$$

depends on the order in which the matrices are multiplied. If norms are taken their product is commutative but, because of the known property of norms, namely

$$\|DG^n(x_0)\| \le \|DG(x_0)\|\|DG(x_1)\| \cdots \|DG(x_{n-1})\|,$$

we have an inequality, that is, the relation

$$\ln\|DG^n(x_0)\| \le \sum_{i=0}^{n-1} \ln\|DG(x_i)\|$$

is not an equality. However, a celebrated theorem by Osledec (1968), known as the multiplicative ergodic theorem provides the necessary extension of ergodic theory to measure-preserving diffeomorphisms. For a diffeomorphism $G : M \hookleftarrow$ preserving a measure μ, and M an m-dimensional manifold, the theorem proves that, putting

$$\lambda(x, w) = \lim_{n\to\infty} \frac{1}{n} \ln\|DG^n(x)w\| \tag{9.15}$$

(where w is a vector in the tangent space at x and $\|w\| = 1$):

(i) the limit (9.15) exists for μ-almost all points $x \in M$;

(ii) as $w \ne 0$ changes in the tangent space, $\lambda(x, w)$ takes $s \le m$ different values;

(iii) if μ is ergodic, the λs are μ-almost everywhere constant, i.e., they do not depend on initial conditions (modulo sets of μ-measure zero). In this case, we can speak of the LCEs of the map G with respect to the G-invariant measure μ.

9.3 Natural, absolutely continuous, SBR measures

To establish that an invariant measure is ergodic may not be enough, however, to make it interesting. In our discussion of dynamical systems from a geometric point of view, we first of all considered the properties of invariance and indecomposability of a set under the action of a map (or a flow). We also tried to establish the conditions under which an invariant set is, in an appropriate sense, observable. In so doing, however, we encountered some conceptual difficulties. In particular, we learned that, if we define attractors as sets to which most points evolve under the action of a flow or a map, they need not be stable (in the sense of Lyapunov) or asymptotically stable sets. We shall now consider the question of observability in the light of ergodic theory.

Consider again the system described by figure 9.1. The stable fixed points $\bar{x}_2$ and $\bar{x}_3$ are observable. If we pick the initial conditions at random and plot the orbit of the system on a computer screen, we have a nonnegligible chance of 'seeing' either of them on the screen (after the transients have died out). On the contrary, the unstable fixed point $\bar{x}_1 = 0$ is not observable, unless we start from points on the y-axis, an unlikely event, *and* there are no errors or disturbances. It is reasonable to conclude that, although all three Dirac measures $\delta_{\bar{x}_1}, \delta_{\bar{x}_2}$ and $\delta_{\bar{x}_3}$ are invariant and ergodic, only the last two are physically relevant.

The basic reason for this fact is that the set defined by 'μ-almost every initial condition' in (9.5) and (9.6) may be too small *vis-à-vis* the state space and therefore negligible in a physical sense. In order to avoid this situation, an invariant measure μ should reflect the asymptotic behaviour of the system not just for μ-almost all points of the state space, but for a physically relevant set, that is, a set of positive k-volume and, therefore, a positive Lebesgue measure. Mathematically, this requirement takes the form

$$\mu = \lim_{n \to \infty} \frac{1}{n} \sum_{i=0}^{n-1} \delta_{G^i(x)} \quad \text{discrete time} \tag{9.16}$$

or

$$\mu = \lim_{\tau \to \infty} \frac{1}{\tau} \int_0^{\tau} \delta_{\phi_t(x)} dt \quad \text{continuous time} \tag{9.17}$$

for $x \in S$ where S is a set of positive Lebesgue measure.[5] Measures satisfying this requirement are sometimes called **natural invariant measures** and the corresponding density (defined below) is called **natural invariant density**; the phrase **physical measure** is also used.

Whereas the properties characterising a natural measure are easily verified in the simpler cases for which the dynamics are not too complicated, in

---

[5] More generally, for any continuous function f, we should have

$$\lim_{n \to \infty} \frac{1}{n} \sum_{i=0}^{n-1} f[G^i(x)] = \int f d\mu$$

for $x \in S$ (and analogously for a flow).

the general case including complex or chaotic systems the determination of the natural invariant measure is a difficult and not entirely solved problem. However, an invariant ergodic measure μ which is absolutely continuous with respect to the Lebesgue measure will automatically satisfy the requirement for a natural measure. To make this clear we introduce the following definition.

definition 9.3 *Given a measurable space $(X, \mathcal{F})$ and two probability measures μ_1 and μ_2, we say that μ_1 is **absolutely continuous** with respect to μ_2 (denoted sometimes as $\mu_1 \ll \mu_2$) if , for any set $B \in \mathcal{F}, [\mu_2(B) = 0] \Rightarrow [\mu_1(B) = 0]$.*

Absolute continuity with respect to the Lebesgue measure m (often called simply **absolute continuity**) of a measure μ thus excludes the possibility that sets which are negligible with respect to the Lebesgue measure (and therefore negligible in a physical sense) are assigned a positive value by μ because, from definition 9.3, it follows that if $\mu \ll m$, then $[\mu(B) > 0] \Rightarrow [m(B) > 0]$. In particular, $\operatorname{supp}(\mu)$ must have positive Lebesgue measure.

If $\mu_1 \ll \mu_2$, there exists a function ρ integrable on X with respect to μ_2, uniquely determined μ_2-almost everywhere, such that $\rho \geq 0$, $\int_X \rho \, d\mu_2 = 1$, and for any measurable set $A \in \mathcal{F}$ we have

$$\mu_1(A) = \int_A \rho(x) \mu_2(dx).$$

In this case, we say that μ_1 is a measure with **density** ρ with respect to μ_2, and we can write $\mu_1 = \rho\mu_2$. Sometimes ρ is written as $d\mu_1/d\mu_2$, the so-called **Radon–Nikodym derivative**. If μ is absolutely continuous (with respect to m), the set defined by μ-almost every initial condition must have positive Lebesgue measure. Therefore, μ reflects the asymptotic behaviour of a physically relevant set of orbits and it qualifies as a 'natural' measure.

Unfortunately, dissipative systems, which form a class of systems often studied in applications, cannot have attractors supporting invariant, absolutely continuous measures. Since, by definition, such systems contract volumes of initial conditions, their attractors must have a (k-dimensional) Lebesgue measure equal to zero.

Because experimental, or computer-generated dynamics of systems seem often to provide a natural selection among the many (even uncountably many) invariant measures, the question arises whether it is possible to define rigorously a class of invariant measures (and thereby a class of dynamical systems) possessing the desired properties of natural measures. An answer to this question is provided by the so-called SBR (Sinai–Bowen–Ruelle) measures, broadly speaking defined as those measures which are absolutely continuous *along the unstable directions*.[6] These are smooth (have densities) in the stretching directions, but are rough (have no densities) in the contracting directions. For systems possessing a SBR measure μ, there exists a subset S of the state space with positive Lebesgue measure such that, for all orbits originating in S, the SBR measure is given by the time averages (9.16) or (9.17).

Although dissipative systems cannot have absolutely continuous invariant measures, they can have SBR measures. SBR measures have been proved to exist for certain classes of systems (e.g., axiom A systems, Anosov diffeomorphisms) and also for the geometric Lorenz attractor (see chapter 6, section 6.8). However, there exist counterexamples showing that they need not exist. For greater detail consider the works mentioned in n. 6, Ruelle (1989) and the bibliography therein.

Another natural way of providing a selection of physical measures is based on the observation that, owing to the presence of noise due to physical disturbances or computer roundoff errors, the time evolution of a system can be looked at as a stochastic process. Under commonly verified assumptions, the latter has a unique stationary measure μ_ε which is a function of the level of noise ε. If this measure tends to a definite limit as the level of noise tends to zero, the limit, which is sometimes called the **Kolmogorov measure**, can be taken as a description of the physical measure. For some, but not all systems the Kolmogorov and the SBR measures actually coincide.

9.4 Attractors as invariant measures

The ideas and results discussed in the previous sections suggest a characterisation of attractors that integrates their geometric and the ergodic features. We present the issue for maps (the corresponding continuous-time version can be obtained easily with slight modifications).

Consider the dynamical system $(X, \mathcal{F}, \mu)$ and the map $G : X \hookleftarrow$ preserving the measure μ. Let us now define the set

$$W = \left\{ x \in X \,\middle|\, \lim_{n \to \infty} \frac{1}{n} \sum_{i=0}^{n-1} f\left[G^i(x)\right] = \int_X f(x)\mu(dx) \right\} \qquad (9.18)$$

where f is any continuous function and write the following definition.

---

[6] A rigorous definition of SBR measures can be found in Eckmann and Ruelle (1985), pp. 639–41. For a somewhat different characterisation of SBR measures, see Keller (1998), pp. 124–30.

definition 9.4 *We say that* **the measure μ is an attractor** *if the set W defined in (9.18) has positive Lebesgue measure.*

Definition 9.4 characterises an attractor as a probability measure on the state space X rather than a subset of that space. The geometric counterpart of the measure-theoretic attractor μ is the smallest closed subset of X with μ-measure one, namely the support of μ. The definition emphasises that, even if $\text{supp}(\mu)$ has zero Lebesgue measure and is therefore nonobservable, for a nonnegligible set of initial conditions the orbits of the system behave in a statistical regular way and the rules governing their long-run behaviour depend on the statistical properties of the attractor, as defined by the invariant measure μ.[7]

9.5 Predictability, entropy

The theoretical apparatus described above will allow us to discuss the question of predictability of chaotic systems in a rigorous manner. In so doing, however, we must remove a possible source of confusion. The ergodic approach analyses dynamical systems by means of probabilistic methods. One might immediately point out that the outcome of deterministic dynamical systems, such as those represented by differential or difference equations, are not random events but, under usually assumed conditions, are uniquely determined by initial values. Consequently, one might conclude, measure and probability theories are not the appropriate tools of analysis. *Prima facie*, this seems to be a convincing argument. If we know the equations of motion of a deterministic system *and we can measure its state with infinite precision*, then there is no randomness involved and the future of the system can be forecast exactly.

Infinite precision of observation is a purely mathematical expression, however, and it has no physical counterpart. When dynamical system theory is applied to real problems, a distinction must therefore be made between *states* of a system, i.e., points in a state space, and *observable states*, i.e., subsets or cells of the state space, whose (nonzero) size reflects our limited power of observation. For example, we cannot verify by observation the statement that the length of an object is π centimetres, a number with an infinitely long string of decimals, thus containing an infinite amount of information. Under normal circumstances, however, we can verify the statement that the length of the object is, say, between 3 and 4 centimetres, or between 3 and 3.2 centimetres and so on. Alternatively, we can think of

---

[7]On this point, cf. Ornstein and Weiss (1991), pp. 79–81; MacKay (1992), p. 5.

the situation occurring when we plot the orbits of a system on the screen
of a computer. What we see are not actual points, but 'pixels' of small
but nonzero size, the smaller the pixels the greater is the resolution of the
graphics environment.

On the other hand, in real systems perfect foresight makes sense only
when it is interpreted as an asymptotic state of affairs which is approached
as observers (e.g., economic agents) accumulate information and learn about
the position of the system. Much of what follows concerns the conditions
under which prediction is possible, given precise knowledge of the equations
of the system, that is, given a deterministic system, but an imprecise, albeit
arbitrarily accurate observation of its state.

The distinction between *state* and *observable state* is unimportant for systems whose orbit structure is simple, e.g., systems characterised by a stable
fixed point or a stable limit cycle. For these systems the assumption of
infinite precision of observation is a convenient simplification and all the interesting results of the investigation still hold qualitatively if that unrealistic
assumption is removed. The distinction, however, is essential for complex,
or chaotic systems. Indeed, one might even say that many of the characterising features of chaotic dynamics, above all their lack of predictability,
can be understood only by taking into account the basic physical fact that
observation can be made arbitrarily but not infinitely precise. We shall
see, later, that the study of the dynamics of observable states provides an
essential link between deterministic and stochastic systems.

After these broad considerations, we now turn to a formal treatment of
the issue. Let us consider a dynamical system $(X, \mathcal{F}, \mu, G)$ where the state
space is restricted to the support of an ergodic measure μ preserved by G.
The idea of finite-precision observation can be given a mathematical form
by defining a **finite partition** $\mathcal{P}$ of X as a collection $\{P_1, P_2, \ldots, P_N\}$ of
disjoint nonempty sets whose union is equal to X. A partition can also be
viewed as an 'observation function' $\mathcal{P} : X \to \{P_1, P_2, \ldots, P_N\}$ such that,
for each point $x \in X$, $\mathcal{P}(x)$ is the element (or cell) of the partition, in
which x is contained. If G is a transformation acting on X, then for any
$\mathcal{P}$, $G^{-1}(\mathcal{P}) = \{G^{-1}(P_i) | P_i \in \mathcal{P}\}$ where $G^{-1}(P_i) = \{x \in X | G(x) \in P_i\}$.
$G^{-1}(\mathcal{P})$ defines another partition of X. Because G is measure-preserving,
then $G^{-1}(\mathcal{P})$ has the same number of elements as $\mathcal{P}$ and corresponding
elements have the same measure.[8]

We also need to introduce the operation called 'joint partition' or **join**,

---

[8]Once again, the reason why we iterate G backward is to deal with noninvertible maps. In
particular, notice that if $\mathcal{P}$ is a partition of X and G is not invertible, $G^i(\mathcal{P})$, $i \geq 1$, need not
be a partition of X.

generally defined as

$$\mathcal{P} \vee \mathcal{Q} = \{P_i \cap Q_j | P_i \in \mathcal{P}, Q_j \in \mathcal{Q}; \ \mu(P_i \cap Q_j) > 0\}$$

where $\mathcal{P}$ and $\mathcal{Q}$ denote two partitions, with elements P_i and Q_j, respectively.

Each observation in a partitioned state space can be regarded as an experiment whose outcome is uncertain. The uncertainty of the experiment, or equivalently the amount of information contained in one observation, can be measured by means of a quantity called the **entropy of a partition** and defined as

$$H(\mathcal{P}) = -\sum_{i=1}^{N} \mu(P_i) \ln \mu(P_i) \tag{9.19}$$

where $\mu(\cup_i P_i) = \mu(X) = 1$. (If we let $\mu(P_i) = 0$ for some i, then we must add the convention $0 \ln 0 = 0$.) Equation (9.19) is the celebrated formula due to Shannon, the founder of the modern theory of information. It satisfies certain basic axioms deemed to be necessary for a 'good' measure of uncertainty, the presentation of which we leave to the appendix.

When dealing with the predictability of a dynamical system we are not interested in the entropy of a partition of the state space, that is, the information contained in a single experiment, but with the entropy of the system as a whole (the rate at which replications of the experiments — i.e., repeated, finite-precision observations of the system as it evolves in time — produce information), when the number of observations becomes very large.

Because the probabilistic structures of the partitions $\mathcal{P}$ and $G^{-1}(\mathcal{P})$ are the same whenever G is measure-preserving, backward iterates of G on $\mathcal{P}$ can be regarded as replications of the same experiment.[9]

The outcome of the 'simple experiment' $\mathcal{P}$ is the observation of the system in one of the elements of $\mathcal{P}$, whereas the outcome of the 'super experiment' defined by the partition

$$\bigvee_{i=0}^{n-1} G^{-i}(\mathcal{P}) \tag{9.20}$$

is the observation of *a sequence of n successive states* occupied by the system under repeated iterations of G, with a finite precision depending on $\mathcal{P}$.

We can evaluate the uncertainty of the super experiment (the amount of information generated by it) by applying the Shannon formula over all of

---

[9] Needless to say, we do not expect, in general, the replications of the experiment to be *independent*.

the elements of (9.20), namely

$$H\left(\bigvee_{i=0}^{n-1} G^{-i}(\mathcal{P})\right). \tag{9.21}$$

If we now divide (9.21) by the number of observations n, we obtain the *average* amount of uncertainty (the average amount of information) of the super experiment. If we then increase the number of observations indefinitely, we obtain[10]

$$h(\mu, \mathcal{P}) = \lim_{n \to \infty} \frac{1}{n} H\left(\bigvee_{i=0}^{n-1} G^{-i}(\mathcal{P})\right) \tag{9.22}$$

which is the *entropy of the system with respect to the partition $\mathcal{P}$*.

The RHS of (9.22) is (the limit of) a fraction. The numerator is the entropy of a partition obtained by iterating G^{-1}, and the denominator is the number of iterations. Intuitively, if, when the number of iterations increases, $H(\bigvee_{i=0}^{n-1} G^{-i}(\mathcal{P}))$ remains bounded, limit (9.22) will be zero; if it grows linearly with n the limit will be a finite, positive value; if it grows more than linearly, the limit will be infinite.[11] To interpret this result, consider that each element of the partition $\bigvee_{i=0}^{n-1} G^{-i}(\mathcal{P})$ corresponds to a sequence of length n of elements of $\mathcal{P}$, that is, an orbit of length n of the system, observed with $\mathcal{P}$-precision. The quantity $H(\bigvee_{i=0}^{n-1} G^{-i}(\mathcal{P}))$ will increase or not with n according to whether, increasing the number of observations, the number of possible sequences also increases.

From this point of view, it is easy to understand why *simple* systems, i.e., those characterised by attractors which are fixed points or periodic orbits, have zero entropy. Transients apart, for those systems the possible sequences of states are limited and their number does not increase with the number of observations. Complex systems are precisely those for which the number of possible sequences of states grows with the number of observations in such a way that limit (9.22) tends to a positive value. For finite-dimensional, deterministic systems characterised by bounded attractors, entropy is bounded above by the sum of the positive Lyapunov exponents and is therefore finite. Quasiperiodic dynamics are an intermediate case for which the number of possible sequences grows linearly ($H(\cdot)$ of equation (9.22) grows *less* than linearly) and therefore, limit (9.22) is zero. We shall return to this case later.

---

[10]For a proof that this limit exists, see Billingsley (1965), pp. 81–2; Mañe (1987), p. 216.

[11]Considering the definition of entropy H, we can see that $H(\bigvee_{i=0}^{n-1} G^{-i}(\mathcal{P}))$ grows linearly if the number of elements of the partition $\bigvee_{i=0}^{n-1} G^{-i}(\mathcal{P})$ (the number of possible sequences of states in the partioned space) grows *exponentially*.

The entropy of a system with respect to a given partition can be given an alternative, very illuminating formulation by making use of the auxiliary concept of **conditional entropy** of $\mathcal{A}$ given $\mathcal{B}$, defined by

$$H(\mathcal{A}|\mathcal{B}) = -\sum_{A,B} \mu(A \cap B) \ln \mu(A|B) \qquad (9.23)$$

where A, B denote elements of the partitions $\mathcal{A}$ and $\mathcal{B}$, respectively. If we, again, think of a partition as an experiment whose outcome is uncertain, then conditional entropy can be viewed as the amount of uncertainty of the experiment $\mathcal{A}$ when the outcome of the experiment $\mathcal{B}$ is known.

By certain calculations that we here omit (cf. Billingsley, 1965, pp. 79–82) it can be shown that

$$h(\mu, \mathcal{P}) = \lim_{n \to \infty} \frac{1}{n} H \left(\bigvee_{i=0}^{n-1} G^{-i}(\mathcal{P}) \right) = \lim_{n \to \infty} H \left(\mathcal{P} \middle| \bigvee_{i=1}^{n} G^{-i}(\mathcal{P}) \right). \qquad (9.24)$$

In light of (9.24), $h(\mu, \mathcal{P})$ can be interpreted as the amount of uncertainty of (the amount of information given by) the $(n+1)$th observation of the system, given the first n observations, in the limit for $n \to \infty$. Zero entropy means that knowledge of the past asymptotically removes all uncertainty and the dynamics of the system are predictable. On the contrary, positive entropy means that, no matter how long we observe the evolution of the system, additional observations still have a positive information content and the system is not entirely predictable.

So far we have been talking about entropy *relative to a specific partition*. The entropy of a system is then defined to be

$$h(\mu) = \sup_{\mathcal{P}} h(\mu, \mathcal{P}) \qquad (9.25)$$

where the supremum is taken over all finite partitions.

The quantity $h(\mu)$ is also known as **Kolmogorov–Sinai (K-S)** or **metric entropy**. Unless we indicate differently, by entropy we mean K-S entropy.

remark 9.6 Metric entropy is a function of the pair (G, μ), but it is commonly (and equivalently) denoted as $h_\mu(G)$, $h_G(\mu)$ (or even $h(G)$ or $h(\mu)$), according to which of the two arguments μ or G one wishes to emphasise.

remark 9.7 If we consider a flow as a one-parameter family of maps, then by 'metric entropy of a flow ϕ_t' we mean the number $h(\phi_1)$, that is, the entropy of the 'time-one map'. On the other hand, for every fixed $t \in \mathbb{R}$, the entropy of the flow map ϕ_t, $h(\phi_t) = |t| h(\phi_1)$.

In the mathematical literature, as well as in applications, one can find a related concept, known as **topological entropy**. Consider a transformation G of the state space X onto itself, together with a partition $\mathcal{P}$ of X. Let $N(\mathcal{P})$ be the number of elements of $\mathcal{P}$. The topological entropy of $\mathcal{P}$ is defined as

$$H_{TOP}(\mathcal{P}) = \ln N(\mathcal{P}). \qquad (9.26)$$

Then the topological entropy of G with respect to $\mathcal{P}$ is:

$$h_{TOP}(G, \mathcal{P}) = \lim_{n \to \infty} \frac{1}{n} H_{TOP} \left(\bigvee_{i=0}^{n-1} G^{-i}(\mathcal{P}) \right). \qquad (9.27)$$

Finally, the **topological entropy** of G is defined as

$$h_{TOP}(G) = \sup_{\mathcal{P}} h_{TOP}(G, \mathcal{P}). \qquad (9.28)$$

Comparing (9.20)–(9.25) and (9.26)–(9.28), the reader will notice that, in the computation of $H_{TOP}(\mathcal{P})$ and consequently of $h_{TOP}(G)$, we have not taken into account the probability of finding the system in each of the cells of the partition $\mathcal{P}$ as quantified by a G-invariant measure μ. Consequently, we have not taken into account the probability of the elements of the 'superpartition' $\bigvee_{i=0}^{n-1} G^{-i}(\mathcal{P})$, that is, of allowed sequences of n states.

If X is compact, there is a simple relation between topological and metric entropies, namely:

$$h_{TOP}(G) = \sup_{\mathcal{M}} h(\mu)$$

where $\mathcal{M}$ is the set of the ergodic measures invariant with respect to G. Hence, positive topological entropy indicates the presence of an invariant ergodic measure and a corresponding invariant set over which the dynamics are chaotic (unpredictable). However, when $h_{TOP}(G) > 0$ but the metric entropy with respect to the natural measure is zero, chaos may take place over a region of the state space which is *too small* to be observed. This phenomenon is nicknamed **thin chaos**.

Actual computation of the metric entropy $h(\mu)$ directly from its definition looks a rather desperate project. Fortunately, a result from Kolmogorov and Sinai guarantees that, under conditions often verified in specific problems, the entropy of a system $h(\mu)$ can be obtained from the computation of its entropy relative to a given partition, $h(\mu, \mathcal{P})$. Formally, we have the following.[12]

---

[12] For discussion and proof of the K-S theorem, see Billingsley (1965), pp. 84–5; Mañe (1987), pp. 218–22.

theorem 9.1 (Kolmogorov–Sinai) *Let $(X, \mathcal{F}, \mu)$ be a measure space; G a transformation preserving μ, and $\mathcal{P}$ a partition of X with finite entropy. If $\vee_{i=0}^{\infty} G^{-i}(\mathcal{P}) = \mathcal{F}$ mod 0, then $h(\mu, \mathcal{P}) = h(\mu)$.*

(In this case mod 0 means: if we exclude certain sets of measure zero.) In this case, $\mathcal{P}$ is called a **generating partition** or a **generator**. Intuitively, a partition $\mathcal{P}$ is a generator if, given a transformation G acting on a state space X, to each point $x \in X$ there corresponds a unique infinite sequence of cells of $\mathcal{P}$, and vice versa. In what follows, we shall repeatedly apply this powerful result.

A simple example will help clarify these ideas. Consider, again, the symmetrical tent map G_Λ on the interval $[0, 1]$ and the partition consisting of the two subintervals located, respectively, to the left and to the right of the $1/2$ point. Thus, we have a partition $\mathcal{P} = \{P_1, P_2\}$ of $[0, 1]$, where $P_1 = \{0 \le x \le 1/2\}$ and $P_2 = \{1/2 < x \le 1\}$. (Recall that measure zero sets do not count so it does not matter where we allocate the point $1/2$.) Then $G_\Lambda^{-1} P_1$ consists of the union of the two subintervals $\{0 < x \le 1/4\}$ and $\{3/4 \le x \le 1\}$ and $G_\Lambda^{-1} P_2$ consists of the union of the two subintervals $\{1/4 < x \le 1/2\}$ and $\{1/2 < x < 3/4\}$. Hence, taking all possible intersections of subintervals, the join $\{G_\Lambda^{-1}(\mathcal{P}) \vee \mathcal{P}\}$ consists of the four subintervals $\{0 \le x \le 1/4\}, \{1/4 < x \le 1/2\}, \{1/2 < x < 3/4\}, \{3/4 \le x \le 1\}$. Repeating the same operation, after $n - 1$ iterates of the map G_Λ^{-1}, the join $\{\vee_{i=0}^{n-1} G_\Lambda^{-i}(\mathcal{P})\}$ is formed by 2^n subintervals of equal length 2^{-n}, defined by $\{x \mid (j - 1)/2^n < x < j/2^n\}, 1 \le j \le 2^n$. Moreover, if we use the Borel σ-algebra the selected partition is a generator for the tent map. Hence, we can apply the Kolmogorov–Sinai theorem and have

$$h(\mu) = h(\mu, \mathcal{P}).$$

Finally, taking into account the fact that the tent map preserves the Lebesgue measure m, we conclude that the K-S entropy of the map is equal to

$$h(m) = \lim_{n \to \infty} \frac{1}{n} H \left(\bigvee_{i=0}^{n-1} G_\Lambda^{-i}(\mathcal{P}) \right) = \lim_{n \to \infty} \frac{1}{n} (-2^n [2^{-n} \ln(2^{-n})]) = \ln 2.$$

Before concluding this section, we would like to recall that entropy is closely linked with another type of statistical invariant, the Lyapunov characteristic exponents. A known result due to Ruelle establishes that for a differentiable map on a finite-dimensional manifold M, preserving an ergodic measure μ with compact support, the following inequality holds

$$h(\mu) \le \sum_{i|\lambda_i>0} \lambda_i, \tag{9.29}$$

where λ denotes an LCE. For systems characterised by a SBR measure, strict equality holds. As we have seen before, the equality indeed holds for the tent map.[13]

The close relation between entropy and LCEs is not surprising. We have already observed that entropy crucially depends on the rate at which the number of new possible sequences of coarse-grained states of the system grows as the number of observations increases. But this rate is strictly related to the rate of divergence of nearby orbits which, in turn, is measured by the LCEs. Thus, the presence of one positive LCE on the attractor signals positive entropy and unpredictability of the system.

9.6 Isomorphism

In the discussion of dynamical systems from a topological point of view, we have encountered the notion of topological conjugacy. Analogously, there exists a fundamental type of equivalence relation between measure-preserving transformations, called **isomorphism** which plays a very important role in ergodic theory and which we shall use in what follows.

definition 9.5 *Let G and $\hat{G}$ be two transformations acting, respectively, on measure spaces $(X, \mathcal{F}, \mu)$ and $(\hat{X}, \hat{\mathcal{F}}, \hat{\mu})$. We say that G and $\hat{G}$ are **isomorphic** (or **metrically isomorphic**) if, excluding perhaps certain sets of measure zero, there exists a one-to-one map θ from X onto $\hat{X}$ such that*

(i) *the following diagram commutes*

$$
\begin{array}{ccc}
X & \xrightarrow{\;G\;} & X \\
\theta \downarrow & & \downarrow \theta \\
\hat{X} & \xrightarrow[\hat{G}]{} & \hat{X}
\end{array}
$$

i.e., we have $\hat{G} \circ \theta = \theta \circ G$;

(ii) *the map θ preserves the probability structure, so that if I and $\hat{I}$ are, respectively, measurable subsets of X and $\hat{X}$, then $m(I) = \hat{m} \circ \theta(I)$ (or $\hat{m}(\hat{I}) = m \circ \theta^{-1}(\hat{I})$).*

---

[13] For technical details, see Ruelle (1989), pp. 71–7; Ornstein and Weiss (1991), pp. 78–85.

Maps θ having the properties (i)–(ii) are called **isomorphisms**.

definition 9.6 *Adopting the same notation as in definition 9.5, suppose the map θ is not one-to-one (and therefore not invertible). If $\hat{G} \circ \theta = \theta \circ G$ and $\hat{m}(\hat{I}) = m \circ \theta^{-1}(\hat{I})$ then we say that $\hat{G}$ is a* **metric factor** *of G.*

Certain properties such as ergodicity and entropy are invariant under isomorphism. Two isomorphic transformations have the same entropy and consequently, if one is chaotic (positive entropy), so is the other. The reverse is true only for a certain class of transformations called Bernoulli which we address in chapter 10, section 10.1. On the other hand, if a dynamical system defined by a map G is ergodic or mixing (see section 9.8) so is any of its (metric) factors.

9.7 Aperiodic and chaotic dynamics

As we mentioned before, it can be easily ascertained that simple systems, characterised by attractors that are fixed points or periodic orbits, all have zero entropy and their dynamics are therefore predictable. We shall then apply the ideas discussed in the previous sections to more complex systems, whose attractors are aperiodic. The asymptotic behaviour of these systems, like that of stochastic processes, can be discussed by considering densities of states, rather than orbits. However, we shall see that complex systems in this sense are not necessarily chaotic in the sense relevant to our present discussion, i.e., they are not necessarily unpredictable. To do so, we shall distinguish between two fundamental classes of behaviour: quasiperiodic (aperiodic but not chaotic) and chaotic.

9.7.1 Quasiperiodic dynamics

Aperiodic nonchaotic (quasiperiodic) behaviour arises in numerous applications in different areas of research. This is often the case when, for certain not unreasonable parameter configurations, a model undergoes a Neimark bifurcation (see chapter 5, section 5.4). Recall that this bifurcation occurs when, owing to a change in the controlling parameter, the modulus of a pair of complex conjugate eigenvalues of the Jacobian matrix calculated at equilibrium becomes greater than one. Excepting certain special resonance cases (see remark 5.5), a Neimark bifurcation generates an invariant (stable or unstable) closed curve the dynamics on which can be represented by a map of the circle and they are periodic or quasiperiodic according to whether the rotation number is rational or irrational. We want to show that, in the quasiperiodic case, entropy is zero and the dynamics are predictable.

Before proving this result we show that, when ρ is irrational, the maps of the circle $G_C(z) = cz$, $c = e^{i2\pi\rho}$ and $G_\rho(x) = x + \rho$ mod 1, (introduced in remark 4.4) are isomorphic and therefore have the same ergodic properties and the same entropy. Consider the following diagram

$$
\begin{array}{ccc}
[0,1) & \xrightarrow{\;G_\rho\;} & [0,1) \\
\theta\downarrow & & \downarrow\theta \\
S^1 & \xrightarrow[\;G_C\;]{} & S^1
\end{array}
$$

where $\theta(x) = e^{i2\pi x}$. The map G_ρ preserves the Lebesgue measure (the pre-image of any subset of $[0,1)$ is an interval of the same length). Analogously, G_C preserves the 'Lebesgue measure on the circle', $\hat{m} = m \circ \theta^{-1}$ (the pre-image under G_C of any arc of S^1 is an arc of the same length). Also we can verify that for $x \in [0,1)$

$$
\theta\left[G_\rho(x)\right] = G_C\left[\theta(x)\right]
$$

or

$$
e^{i2\pi(x+\rho)} = e^{i2\pi\rho}e^{i2\pi x}.
$$

Thus the diagram commutes. Moreover, by the definition of $\hat{m}$ it is easy to see that for any measurable subset I of $[0,1)$ we have

$$
m(I) = \hat{m}\left[\theta(I)\right]
$$

or, for any measurable subset of S^1

$$
\hat{m}(\hat{I}) = m\left[\theta^{-1}(\hat{I})\right].
$$

Therefore, G_C and G_ρ are isomorphic.

We can now use two standard results of ergodic theory and state the following:

proposition 9.1 *In the irrational case, the rotation map G_C is ergodic (with respect to the invariant measure $\hat{m}$).*

For a proof see, for example, Doob (1994), pp. 120–1.

proposition 9.2 *In the irrational case, the map G_C has zero entropy.*

PROOF[14] First, consider that the partition $\mathcal{P}_N$ of the circle into N equal intervals is a generator and thus, for the map G_C preserving the Lebesgue measure on the circle $\hat{m}$, we have

$$h(\hat{m}, \mathcal{P}_N) = h(\hat{m}).$$

Next, consider that if ρ is irrational, the joint partition

$$H\left(\bigvee_{i=0}^{n-1} G_C^{-i}(\mathcal{P}_N)\right)$$

contains Nn equal elements and therefore, for the map G_C

$$h(\hat{m}) = \lim_{n\to\infty} \frac{1}{n} \ln(Nn) = 0.$$

Then, because of isomorphism, we have the following.

proposition 9.3 *If ρ is irrational, the map G_ρ is ergodic (with respect to the Lebesgue measure) and has zero entropy.*

9.7.2 Chaotic dynamics

Let us take first the logistic map

$$G(x) = \mu x(1-x) \qquad 1 \le \mu \le 4. \tag{9.30}$$

Map (9.30) and other analogous one-dimensional, 'one-hump' maps have often been used as mathematical idealisations of problems arising in many different fields. It is known that for system (9.30) there exists a nonnegligible (positive Lebesgue measure) set of values of the parameter μ for which its dynamics are chaotic.[15] Recall that in chapter 6, section 6.5 we studied the behaviour of the logistic map for $\mu = 4$, denoted G_4, by way of conjugacy with the even simpler tent map G_Λ. The tent map is one of the very few maps for which exact results are available and we used it to calculate the Lyapunov characteristic exponent for G_4 exactly. We can now show that the maps G_4 and G_Λ are isomorphic, and therefore have the same metric entropy. Condition (i) of definition 9.5 has already been verified for maps G_4 and G_Λ in section 6.5 for $\theta(x) = \sin^2(\pi x/2)$. Condition (ii) follows from the result established in example 6 at the end of section 9.1 and, in particular, from the fact that G_4 preserves the measure $m \circ \theta^{-1}$ where m is the Lebesgue measure. Therefore, we can write the following.

---

[14]Cf. Mañe (1987), p. 222; Katok and Hasselblatt (1995), pp. 173–4.
[15]Cf. Jakobson (2000), p. 242, theorem 2.2.

proposition 9.4 *The logistic map with parameter $\mu = 4$, G_4, and the tent map G_Λ, are isomorphic.*

Because isomorphism preserves entropy, we can conclude that the logistic map has entropy equal to $\ln 2 > 0$ and its dynamics are therefore unpredictable. Notice that, in this case, the metric entropy and the unique Lyapunov characteristic exponent (for most initial conditions) are equal.

Consider now the doubling map G_D on the interval

$$G_D : [0,1] \to [0,1] \qquad G_D(x) = \begin{cases} 2x, & \text{if } 0 \le x \le \frac{1}{2}, \\ 2x - 1, & \text{if } \frac{1}{2} < x \le 1 \end{cases}$$

and the following diagram

$$
\begin{array}{ccc}
[0,1] & \xrightarrow{\;G_D\;} & [0,1] \\
{\scriptstyle \theta}\downarrow & & \downarrow{\scriptstyle \theta} \\
[0,1] & \xrightarrow[\;G_\Lambda\;]{} & [0,1]
\end{array}
$$

where G_Λ is the tent map. We already know that the map G_Λ preserves the Lebesgue measure and it is easy to verify that the map G_D does as well.

Let $\theta = G_\Lambda$. (Notice that G_Λ plays two distinct roles in the diagram: it is one of two maps investigated and it is also the map connecting them.) G_Λ is, of course, two-to-one and it is not an isomorphism between G_D and G_Λ. However,

 (i) the diagram commutes, that is, we have

$$G_\Lambda[G_D(x)] = G_\Lambda[G_\Lambda(x)]$$

for $x \in [0,1]$;
 (ii) G_Λ is measure-preserving, in the sense that, for any measurable subset $I \subset [0,1]$, we have

$$m(I) = m\left[G_\Lambda^{-1}(I)\right]$$

(in view of the fact that m is G_Λ-invariant).

Therefore G_Λ is a metric factor of G_D.

Consider now, the following proposition (cf. Katok and Hasselblatt, 1995, pp. 171–2, proposition 4.3.16).

proposition 9.5 *If the map $\hat{G}$ is a metric factor of G, according to definitions 9.5 and 9.6, then $h_{\hat{\mu}}(\hat{G}) \le h_\mu(G)$ where, in general, $h_\mu(T)$ is the metric entropy of the map T with respect to the T-invariant probability measure μ.*

Therefore, because we know that $h_m(G_\Lambda) = \ln 2 > 0$, it must also be true that $h_m(G_D) \geq h_m(G_\Lambda) > 0$.

9.8 Mixing

For dynamical systems defined by measure-preserving transformations, there exists another characterisation of chaos called *mixing* that focuses on the probabilistic properties of orbits.

definition 9.7 *A dynamical system* $(X, \mathcal{F}, \mu, G)$ *is said to have the mixing property or, for short, to be* **mixing***, if*

(a) for every pair $A, B \in \mathcal{F}$

$$\lim_{n \to \infty} \mu\left[A \cap G^{-n}(B)\right] = \mu(A)\mu(B); \qquad (9.31)$$

(b) or, equivalently, for any two square integrable functions f *and* g

$$\lim_{n \to \infty} \int_X f[G^n(x)]g(x)d\mu = \int_X f(x)d\mu \int_X g(x)d\mu. \qquad (9.32)$$

Sometimes we describe this situation by saying that 'the transformation G *is mixing'.*

An intuitive, physical interpretation of mixing is the following. Consider a cocktail shaker S full of, say, 90% cola and 10% rum, and let the transformation G denote the action of shaking. If the liquid is incompressible, we can assume that G preserves the three-dimensional Lebesgue measure, m. If R is the part of S originally occupied by rum, then the relative amount of rum contained in any other part A of the shaker, after k shakes is given by

$$\frac{m\left[G^{-k}(A) \cap R\right]}{m(A)}. \qquad (9.33)$$

If G is mixing, (9.33) tends to $m(R)$ as $k \to \infty$. Metaphorically speaking, if the barman is good at shaking cocktails, then after a sufficiently large number of shakes, any portion of the shaker contents, small as you wish, should contain approximately 10% of rum.

We can also look at mixing in a different, but equivalent, way by considering the **correlation function** of the map G defined as

$$\mathrm{cor}_G(f, g; n) = \left| \int_X f[G^n(x)]g(x)\,d\mu - \int_X f(x)\,d\mu \int_X g(x)\,d\mu \right|,$$

where, again, f and g are any two square integrable functions and μ is a G-invariant measure. If G is mixing, from (9.31) and (9.32) we deduce that

$$\lim_{n \to \infty} \text{cor}_G(f, g; n) = 0.$$

Thus, the mixing property implies asymptotic vanishing of correlation.

The invariant measure μ of a mixing system has an interesting stability property (cf. Cornfeld *et al.*, 1982, pp. 22–6). Let μ_0 be any arbitrary normalised measure, absolutely continuous with respect to μ (see definition 9.3) and let ρ be the Radon–Nikodym derivative $d\mu_0/d\mu$. For any measurable set A, let us define

$$\mu_n(A) = \mu_0 \left[G^{-n}(A) \right]$$

or

$$\mu_n(A) = \int_X \chi_A \left[G^n(x) \right] d\mu_0 = \int_X \chi_A \left[G^n(x) \right] \rho(x) \, d\mu$$

(where χ_A is the indicator function for A).

If G is mixing, because of (9.32)

$$\lim_{n \to \infty} \int_X \chi_A \left[G^n(x) \right] \rho(x) \, d\mu = \int_X \chi_A(x) \, d\mu \int_X \rho(x) \, d\mu.$$

But the first integral on the RHS is equal to $\mu(A)$ and the second is equal to one. Consequently,

$$\lim_{n \to \infty} \mu_n(A) = \mu(A).$$

Thus, for a mixing system, initial arbitrary probability distributions converge asymptotically to the invariant one.

remark 9.8 A few last points are in order:

(1) There exists a weaker characterisation of mixing, called **weakly mixing**, for which the property (9.32) is replaced by the following

$$\lim_{n \to \infty} \frac{1}{n} \sum_{k=0}^{n-1} \left(\int_X f[G^k(x)]g(x) \, d\mu - \int_X f(x) \, d\mu \int_X g(x) \, d\mu \right)^2 = 0.$$

(2) The notion of mixing (and weakly mixing) can be extended to flows with slight modifications in the relevant definitions.

(3) An interesting question is whether, and under what conditions, different characterisations of chaos are equivalent. This problem is discussed in detail in Oono and Osikawa (1980). The authors find that for *endomorphisms of the interval*, all the common characterisations of chaos (including mixing, positive entropy and positive Lyapunov exponent) are

equivalent. This result covers the simple, one-dimensional examples discussed in the previous sections but does not extend to multi-dimensional systems for which there exist counterexamples such as mixing systems with zero LCE and systems with positive LCE but not mixing.

Appendix: Shannon's entropy and Khinchin's axioms

Shannon's entropy as defined by equation (9.19) satisfies the following four axioms suggested by Khinchin as necessary requirements for an acceptable measure of information (or uncertainty)H:

(i) H should be a function of the probabilities p_i only; i.e., it should be $H(p)$, $p \equiv (p_1, p_2, \ldots, p_N)$.

(ii) H should take its maximum value when the distribution of probability is uniform, i.e., when we have $p_i = 1/N$ for all i, namely:

$$\arg \max H(p) = \underbrace{\frac{1}{N}, \frac{1}{N}, \ldots, \frac{1}{N}}_{N \text{ times}}.$$

(iii) The value of $H(p)$ should remain unchanged if we add to the sample one additional event with probability zero, $H(p; 0) = H(p)$.

(iv) Suppose there are two random variables ξ_1 and ξ_2 with ξ_1 taking N_1 values with probability $p_i^1 (i = 1, 2, \ldots, N_1)$ and ξ_2 taking N_2 values with probability $p_j^2 (j = 1, 2, \ldots, N_2)$ and consider the compound system with microstates (i, j). In general, the probability of the state (i, j) of the combined system will be given by

$$p_{i,j} = Q(j|i)p_i^1$$

where $Q(j|i)$ is the conditional probability that subsystem 2 is in state j if subsystem 1 is in state i. The fourth axiom requires that the

$$H(p) = H(p^1) + \sum_i p_i^1 H(Q|i)$$

where

$$H(Q|i) = -\sum_j Q(j|i) \ln Q(j|i).$$

In the case for which the two subsystems are statistically independent, $Q(j|i) = p_j^2$ and, therefore, $p_{ij} = p_i^1 p_j^2$, and consequently, for the combined system, $H(p) = H(p^1) + H(p^2)$. In the general case, the fourth axiom requires that the evaluation of information (or of uncertainty) should be independent of the way it is collected, whether simultaneously from the whole system or successively from the subsystems.[16]

Exercises

9.1   Prove equation (9.14).

9.2   Show that for the system discussed in example 2 (see figure 9.1), the property defined by (9.17) is verified for the invariant measures $\delta_{\bar{x}_2}$ and $\delta_{\bar{x}_3}$, but not $\delta_{\bar{x}_1}$.

9.3   Prove the statement contained in example 2 (μ is not ergodic).

9.4   Let G_Λ be the tent map:

$$G : [0,1] \to [0,1]$$

$$G_\Lambda(y) = \begin{cases} 2y, & \text{if } 0 \le y \le \frac{1}{2}, \\ 2(1-y), & \text{if } \frac{1}{2} < y \le 1 \end{cases}.$$

Show that it preserves the Lebesgue measure m. Consider now the space of one-sided infinite sequences, defined as follows:

$$M = \{\{x^i\}_0^\infty \mid x^i \in [0,1], \ G_\Lambda x^i = x^{i-1}, \ i > 0\}.$$

Let T be the map

$$T : M \to M$$

$$T(\{x^i\}) = \{Tx^i\} \quad i \ge 0$$

and μ a probability measure on M defined by

$$\mu(A) = m(I)$$

where

$$A = \{\{x^i\}_0^\infty \in M \mid x^k \in I \subset [0,1], \text{ for a fixed } k \ge 0\}$$

(that is, A is the set of infinite sequences in M such that its kth element belongs to the subset $I \subset [0,1]$ and the measure μ assigns to A a value equal to the Lebesgue measure of I).

(a) Show that T is invertible and preserves μ.

---

[16] For further discussion of this point and a proof that Shannon's entropy is *uniquely* determined by the four axioms, see Beck and Schlögl (1993), pp. 47–9.

(b) Show that

$$\mu\left(\{\{x^i\}_0^\infty \in M \mid x^0 \in I_0, \dots, x^r \in I_r\}\right)$$
$$= m\left(G_\Lambda^{-r}(I_0) \cap G_\Lambda^{-r+1}(I_1) \cap \dots \cap I_r\right)$$

where $I_i (i = 0, \dots, r)$ are subsets of $[0, 1]$.

9.5   Consider the map of the unit circle (multiplicative notation)

$$G : S^1 \to S^1 \qquad G(z) = z^2.$$

Show that the map G preserves the circular Lebesgue measure.

9.6   Suppose the logistic map on the interval $[0, 1]$ has a periodic orbit of period k. Prove that the measure that assigns the value $\frac{1}{k}$ to each of the k periodic points is invariant with respect to G.

9.7   Consider the map of the interval

$$G : [0, 1) \to [0, 1)$$
$$G(x) = \begin{cases} 2x_n, & \text{if } 0 \le x < \frac{1}{2} \\ 2x_n - 1, & \text{if } \frac{1}{2} \le x < 1 \end{cases}$$

and show that it can be used as a mathematical description of the stochastic process generated by repeated tossing of a coin.

9.8   Consider again the doubling map on the interval $[0, 1)$. Suppose that the state of the system at the initial time can be observed with finite precision, say $x_0 = 0.15 \pm 0.05$. After how many iterates of G will the information contained in the initial observation be lost entirely? Try to relate your answer to the LCE of the map.

9.9   Consider the logistic map G with $\mu = 2$. What is the limit set of a generic initial point? Can you find an initial point x_0 such that the LCE for the orbit of x_0, $\lambda(x_0) = \ln 2$? What is the G-invariant probability measure μ such that this result holds for μ-almost all initial points?

9.10 Show that the distance function

$$d(z_1, z_2) = |z_1 - z_2|$$

(where $z \in \mathbb{C}$, $|z| = 1$, $z = e^{i2\pi\theta}$, $\theta \in \mathbb{R}$) is a metric on S^1.

9.11 Recalling example 6 and the solution to exercise 6.7, find a probability measure invariant with respect to the map $F : [-1, 1] \to [-1, 1]$, $F(x) = 2x^2 - 1$ and show that the map F is isomorphic to the tent map.

10

Deterministic systems and stochastic processes

In chapter 9 we discussed the question of predictability by means of the concept of entropy. As we shall see, however, not all systems with positive entropy are equally unpredictable. Whereas zero entropy implies that the dynamics of a system are predictable with regards to *any* possible finite partition, positive entropy simply means that the system is unpredictable with regards to *at least one* partition. However, there exist systems that are unpredictable for *any* possible partition. Among the latter there exists a special class, called Bernoulli, which is the most chaotic, or the least predictable of all, and to which we now turn our attention.

10.1 Bernoulli dynamics

The first step in making the above informal statements more precise is to provide a unified characterisation of **abstract dynamical systems**. This will allow us to discuss the relations between deterministic chaotic systems and random, or stochastic processes more rigorously and effectively.

Let Ω be the set of all bi-infinite sequences $\omega = (\ldots, x_{-1}, x_0, x_1, \ldots)$ where the coordinate x_i of each element of Ω are points of a certain measurable space $(X, \mathcal{F})$. A natural σ-algebra $\tilde{\mathcal{F}}$ for the space Ω is defined in terms of **cylinder sets** C, thus:

$$C = \{\omega = (\ldots, x_{-1}, x_0, x_1, \ldots) \in \Omega \mid x_{i_1} \in C_1, \ldots, x_{i_k} \in C_k\}$$

where $C_1, \ldots, C_k \in \mathcal{F}$. Suppose now that μ is a measure on $\tilde{\mathcal{F}}$, normalised so that $\mu(\Omega) = 1.$[1] In the language of probability theory, the triplet $(\Omega, \tilde{\mathcal{F}}, \mu)$ is called a (discrete-time) **stochastic process**, taking values in the state space $(X, \mathcal{F})$. If the measure $\mu(\{\omega \in \Omega \mid x_{i_1+n} \in C_1, x_{i_2+n} \in C_2, \ldots, x_{i_k+n} \in C_k\})$

---

[1] The measure space $(\Omega, \tilde{\mathcal{F}}, \mu)$ may have to be 'completed' so that subsets of measure zero sets may be measurable (with zero measure). See, for example, Billingsley (1979), p. 35.

does not depend on n, $-\infty < n < \infty$, we say that the process is **stationary**. In this case, the measure μ is invariant with respect to the shift map $T :$ $\Omega \to \Omega$ defined as

$$\omega' = T(\omega), \quad \omega' = (\ldots, x'_{-1}, x'_0, x'_1, \ldots) \quad x'_i = x_{i+1}, \quad i \in \mathbb{Z}.$$

In particular, for any measurable set $C \in \tilde{\mathcal{F}}$, and any $-\infty < n < \infty$, $\mu\left[T^{-n}(C)\right] = \mu(C)$. Therefore, the quadruplet $(\Omega, \tilde{\mathcal{F}}, \mu, T)$ defines a dynamical system, and T is an automorphism of the space $(\Omega, \tilde{\mathcal{F}}, \mu)$.

An analogous argument can be developed for the set Ω_+ of one-sided infinite sequences. In this case we use a one-sided shift T_+, that is, a shift that moves a sequence one step to the left and drops the first element. In this case, T_+ is an endomorphism of the space $(\Omega_+, \tilde{\mathcal{F}}_+, \mu_+)$.

remark 10.1 When deriving an abstract dynamical system from a (stationary) stochastic process, care must be used to distinguish the *state space of the dynamical system* (in the present notation, the space of infinite sequences Ω), on which T acts, and the *state space of the process* (in this case the space X).

In probability theory, a special role is played by stationary processes which are 'completely random' in the sense that they describe sequences of independent trials[2] with the same probability of observation ρ (or the same probability density in the case of a continuous sample space). In this case, the state space of the dynamical system is a Cartesian product

$$\left(\Omega, \tilde{\mathcal{F}}, \mu\right) = \prod_{n=-\infty}^{\infty} \left(X^{(n)}, \mathcal{F}^{(n)}, \rho^{(n)}\right)$$

where $\left(X^{(n)}, \mathcal{F}^{(n)}, \rho^{(n)}\right) = (X, \mathcal{F}, \rho)$ for all n. The measure μ is the **product measure** generated by the measure ρ, sometimes denoted by $\mu = \prod_n \rho^{(n)}$.

These processes are called **Bernoulli** after the famous Swiss probabilist. Then T, the two-sided shift defined as before, is called a **Bernoulli automorphism** or **Bernoulli shift** on Ω. In an analogous manner, we can define a **one-sided Bernoulli shift**, or a **Bernoulli endomorphism** T_+ on the space of one-sided sequences Ω_+.

The simplest example of an experiment whose representation is a Bernoulli process is given by the repeated tossing of a fair coin, with $1/2$

---

[2]Recall that if $x_{i_1}, x_{i_2}, \ldots, x_{i_k}$ is a sequence of random variables with state space $(X, \mathcal{F})$, we say that the x_{i_j} are **independent** if for any sequence $C_1, C_2, \ldots, C_k \in \mathcal{F}$, the events $\{x_{i_1} \in C_1\}, \{x_{i_2} \in C_2\}, \ldots, \{x_{i_k} \in C_k\}$ are independent in the sense that $\mathrm{prob}(x_{i_1} \in C_1, x_{i_2} \in C_2, \ldots, x_{i_k} \in C_k) = \mathrm{prob}(x_{i_1} \in C_1) \cdot \mathrm{prob}(x_{i_2} \in C_2) \cdots \mathrm{prob}(x_{i_k} \in C_k)$.

probability of heads and $1/2$ probability of tails. Then the probability
that an infinite sequence of tosses of the coin contains any *given* finite k-
subsequence of heads and tails is equal to $(1/2)^k$. The process is commonly
denoted by $B(1/2, 1/2)$. On the other hand, the Bernoulli shift $B(p_1, \ldots, p_k)$
is the mathematical description of a process consisting of spinning a roulette
wheel with k slots of widths p_i, $p_i \geq 0$ and $\sum_i p_i = 1$.

definition 10.1 *A process is said to have the Bernoulli property or, for
short, to be a* **B-process***, if it is isomorphic to a Bernoulli process. A flow
$\phi_t(x)$ is said to be a B-process if and only if the time-one map $\phi_1(x)$ is.*

B-processes include both deterministic (discrete- or continuous-time) dynamical systems and stochastic processes. Examples of the former include
the logistic map (for $\mu = 4$), the tent map, the Lorenz geometric model,
the Shilnikov model; examples of the latter include i.i.d. processes and
continuous-time Markov processes.

B-processes are characterised by a property called 'Very Weak Bernoulli
(VWB)'. A rigorous definition of this property is beyond the scope of this
chapter (see, Ornstein (1974), pp. 41–4 and Ornstein and Weiss (1991), pp.
58–60) and we provide only an intuitive description. Broadly speaking, a
VWB is an asymptotic form of independence. For an independent process
the probability of a future event, conditioned on *any* past event is the same
as the unconditional probability. For a VWB process the future, conditional on the past, converges in time (in an appropriately defined metric) to
the unconditional future. In other words, for B-processes, the information
provided by the past regarding the long-term future becomes vanishingly
small.

The crucial role played by Bernoulli dynamical systems in chaos theory
can be better appreciated in the light of two well-known basic results. (For
proofs see Walters (1982), p. 108; Cornfeld and Sinai (2000), p. 50.)

proposition 10.1 *If G is an automorphism on the measure space $(X, \mathcal{F}, \mu)$
which is chaotic in the sense that the metric entropy $h(G) > 0$, and if G_1 is
a Bernoulli automorphism with $h(G_1) \leq h(G)$, then there exists a measure-
preserving transformation θ such that $\theta \circ G = G_1 \circ \theta$, that is, G_1 is a (metric)
factor of G.*

proposition 10.2 *Any automorphism G_1 which is a factor of a Bernoulli
automorphism G is also a Bernoulli automorphism.*

These two results indicate that although not all chaotic (positive entropy)
systems are Bernoulli, there is a 'Bernoulli core' in each of them.[3]

---

[3]There exists an intermediate class of invertible transformations which are known as **Kol-**

We now use these general ideas and results to show that two simple maps on the interval, which we have discussed previously and which are quite common in applications, are indeed not only chaotic but Bernoulli.

proposition 10.3 *The symmetrical tent map* $G_\Lambda : [0,1] \to [0,1]$

$$G_\Lambda(x) = \begin{cases} 2x, & \text{if } 0 \le x \le \frac{1}{2}, \\ 2(1-x), & \text{if } \frac{1}{2} < x \le 1. \end{cases}$$

is isomorphic (modulo 0) to the one-sided Bernoulli shift T_+ *on the two-symbol space* Σ_{2+}.

PROOF Consider the two subintervals $I_- = [0, 1/2)$ and $I_+ = (1/2, 1]$ and the set $I_0 = \{x \in [0,1] \mid G_\Lambda^n(x) \in I_+ \text{ or } I_- \ \forall n \ge 0\}$, that is, the set of points of the interval which are never mapped to its middle point $1/2$. Let Σ_{2+} denote the space of one-sided infinite sequences of the symbols $0, 1$ (i.e., $\Sigma_{2+} = \{0,1\}^{\mathbb{Z}_+}$), and define the map $\theta : I_0 \to \Sigma_{2+}$

$$\theta(x) = s \equiv (s_0 s_1 \ldots s_n \ldots)$$

where

$$s_k = \begin{cases} 0, & \text{if } G_\Lambda^k(x) \in I_-, \\ 1, & \text{if } G_\Lambda^k(x) \in I_+. \end{cases}$$

Notice that the map θ is one-to-one, 'nearly onto', because the set $\Sigma_{2+} \setminus \theta(I_0)$ contains countably many points of the form $s = (s_0 s_1 \ldots s_k \ 1 \ 0 \ldots \{0\}^\infty)$, and that θ is a homeomorphism between I_0 and $\theta(I_0)$. Moreover, if T_+ denotes again the one-sided shift map on Σ_{2+}, we have $T_+ \circ \theta = \theta \circ G_\Lambda$. Let μ_1 be a (stationary) probability measure on Σ_{2+} defined by

$$\mu_1(C) = 2^{-k}$$

where C is any cylinder set

$$C = \{s \in \Sigma_{2+} \mid s_{i_1} = c_1, \ldots, s_{i_k} = c_k\}$$

where $c_j (j = 1, 2, \ldots, k)$ are fixed values equal to 0 or 1, and for any (i, j)

$$\text{prob}(s_{i_j} = 1) = \text{prob}(s_{i_j} = 0) = \frac{1}{2}.$$

We shall denote this measure by $\mu_1(1/2, 1/2)$. The measure μ_1 is invariant

mogorov, or simply **K-automorphisms**, which are unpredictable for any possible partition. Although all Bernoulli automorphisms are K, the reverse need not be true (cf. Ornstein and Weiss 1991, pp. 19–20). K-automorphisms are seldom met in applications and we shall not deal with them here. There exists an analogous class of noninvertible transformations (endomorphisms) called **exact** (cf. Cornfeld *et al.* 1982, p. 289).

with respect to T_+. Notice also that, because $\Sigma_{2+} \setminus \theta(I_0)$ is a countable set, $\mu_1[\theta(I_0)] = 1$ and therefore $\Sigma_{2+} \setminus \theta(I_0)$ is μ_1-negligible. In order to prove isomorphism, we look for a probability measure μ_2 on $[0,1]$, invariant with respect to G_Λ and such that for any measurable subset $C \subset \Sigma_{2+}$ we have

$$\mu_1(C) = \mu_2\left[\theta^{-1}(C)\right]$$

or for any measurable subset $A \subset [0,1]$

$$\mu_2(A) = \mu_1[\theta(A)] = \mu_1\left[\theta(A \bigcap I_0)\right].$$

We can verify that the Lebesgue measure m which we know to be G_Λ-invariant, satisfies this condition. To see this, consider that for any 'cylinder set' C defined as before, the set $\theta^{-1}(C)$ is a subinterval of $[0,1]$ which is mapped onto $[0,1]$ by a linear branch of G_Λ^k with slope equal to 2^k. Thus, the length of an interval $\theta^{-1}(C)$, and therefore its Lebesgue measure, is equal to 2^{-k}. Hence, setting $\mu_2 = m$, we have the desired result

$$\mu_1(C) = m\left[\theta^{-1}(C)\right].$$

remark 10.2 The above result can be strengthened and generalised in various directions. It can be shown that

(i) the isomorphic relation with a Bernoulli shift exists for a generalised, not necessarily symmetric, tent map;

(ii) the probability measure $\mu_2 = m$ is the only invariant measure on $[0,1]$ which satisfies the aforementioned conditions and is absolutely continuous with respect to the Lebesgue measure;

(iii) an extension of the state space (call it $[0,1]^+$) and a corresponding modification of the map G_Λ (call it G_Λ^+) can be introduced such that the map θ is a homeomorphism $\theta : [0,1]^+ \to \Sigma_{2+}$ and that the action of G_Λ^+ on $[0,1]^+$ is equivalent from a measure-theoretic point of view to that of G_Λ on $[0,1]$.

For technical details cf. Keller (1998), pp. 16–17 and 166–7, on which the present discussion is based.

proposition 10.4 *The logistic map (see section 6.5)*

$$G_4 : [0,1] \to [0,1] \qquad G_4(x) = 4x(1-x)$$

is isomorphic to the tent map. Therefore the map G_4 is Bernoulli.

PROOF Isomorphism is proved in proposition 9.4. The rest follows from the definitions.

proposition 10.5 *The doubling map on the interval*

$$G_D : [0,1] \to [0,1] \qquad G_D(x) = \begin{cases} 2x, & \text{if } 0 \le x \le \frac{1}{2}, \\ 2x - 1, & \text{if } \frac{1}{2} < x \le 1 \end{cases}$$

is isomorphic (modulo 0) to the one-sided Bernoulli shift on two symbols and therefore, it is Bernoulli.

PROOF The proof is similar to that of proposition 10.3 and we abbreviate it. Let $(\Sigma_{2+}, T_+, \mu(1/2, 1/2))$ where $\mu(1/2, 1/2)$ is the same as μ_1 defined in the proof of proposition 10.3. Let $\mathcal{P} = \{P_0, P_1\}$ be a partition of $[0,1]$ with $P_0 = [0, 1/2]$ and $P_1 = (1/2, 1]$. The map $\theta : \Sigma_{2+} \to [0,1]$

$$\theta(s) = \bigcap_{j=0}^{\infty} G_D^{-j} P_{s_j}$$

(where $s_j = 0$ or 1) is one-to-one and invertible (modulo sets of measure zero) and the diagram

$$
\begin{array}{ccc}
\Sigma_{2+} & \xrightarrow{\;T_+\;} & \Sigma_{2+} \\
{\scriptstyle \theta(s)}\big\downarrow & & \big\downarrow{\scriptstyle \theta(s)} \\
[0,1] & \xrightarrow[\;G_D\;]{} & [0,1]
\end{array}
$$

commutes.

Consider now that G_D preserves the Lebesgue measure m, because the pre-image of any measurable subset I of $[0,1]$ is the union of two subsets, the length of which is equal to half that of I. For a given sequence $\{s_j\}_0^{k-1}$, $s_j \in \{0, 1\}$, the length of the subset $I^{(k)} \subset [0,1]$

$$I^{(k)} = \bigcap_{j=0}^{k-1} G_D^{-j} P_{s_j}$$

is equal to 2^{-k}. Also, if $C^{(k)}$ is the cylinder set defined by the same sequence, we have $\mu(C^{(k)}) = 2^{-k}$. Because $I^{(k)} = \theta(C^{(k)})$, it follows that $\mu(C^{(k)}) = m[\theta(C^{(k)})]$ (and $m(I^{(k)}) = \mu[\theta^{-1}(I^{(k)})]$). We can conclude that θ is an isomorphism between the shift map T_+ and the map G_D. Thus, G_D is Bernoulli.

10.2 Markov shifts

A second, basic example of dynamical systems arising in probability theory
is provided by Markov shifts (cf. Cornfeld *et al.* 1982, pp. 182–8). Broadly
speaking, we say that a stochastic process is a **Markov process** (or **chain**)
if the future of the process is independent of the past, given its present
value (one might say that the process has short memory). In order to
provide a more rigorous characterisation of this idea, let us consider again a
stochastic process $(\Omega, \tilde{\mathcal{F}}, \mu)$ taking values on a state space $(X, \mathcal{F})$. Then, the
evolution of a Markov process in time can be described in terms of a one-step
transition probability kernel (also called **stochastic operator**)

$$P(x, A) \quad x \in X \quad A \in \mathcal{F}$$

where

(i) for each A, $P(\cdot, A)$ is a measurable function on X; and
(ii) for each x, $P(x, \cdot)$ is a probability measure on $\mathcal{F}$.

We say that a probability measure ρ on $\mathcal{F}$ is invariant with respect to P,
or equivalently that P preserves ρ, if we have

$$\rho(A) = \int_X P(x, A)\rho(dx).$$

A transition probability kernel P and a corresponding invariant measure ρ
define a measure μ on the space of sequences Ω as follows:

$$\mu(C) = \int_{C_1} \rho(dx_i) \int_{C_2} P(x_i, dx_{i+1}) \cdots \int_{C_r} P(x_{i+r-2}, dx_{i+r-1})$$

where again

$$C = \{\omega \in \Omega | x_i \in C_1, \ldots, x_{i+r-1} \in C_r\} \in \tilde{\mathcal{F}}$$

$\omega = (\ldots, x_{-1}, x_0, x_1, \ldots)$ with $-\infty < i < \infty$ and $C_i \in \mathcal{F}$. Invariance of ρ
(with respect to P) implies stationarity of the process (that is, given P and
ρ, μ does not depend on the choice of i).

The shift map T on the measure space $(\Omega, \tilde{\mathcal{F}}, \mu)$ is a dynamical system
called a **Markov shift** (or **Markov automorphism**). In a perfectly analogous manner it is possible to define Markov shifts on the space Ω_+ of
one-sided infinite sequences, i.e., **Markov endomorphisms**.

A special, interesting case occurs when the state space X of the stochastic
process is a finite set, that is, we have $X = \{X_1, \ldots, X_m\}$. In that case the
transition probability kernel reduces to a transition matrix $\Pi = (\pi_{ij})$ where
$\pi_{ij} = P(X_i, \{X_j\})$, $i, j = 1, \ldots, m$. The measure ρ invariant with respect

to P is now defined by a vector $\pi^T = (\pi_1, \ldots, \pi_m)$ where $\pi_k = \rho(\{X_k\})$ for $k = 1, \ldots, m$ and $\pi^T = \pi^T \Pi$.

The matrix Π is said to be a **stochastic matrix** because $\pi_{ij} \geq 0$, $\sum_{j=1}^{m} \pi_{ij} = 1$. Notice that the Bernoulli shift is a special case of a Markov shift with $P(x, C) = \rho(C)$ or, in the case of a finite state space, $\pi_{ij} = \pi_j$.

Let $\Pi^{(n)} = (\pi_{ij}^{(n)})$ denote the nth iterate of the matrix Π. The Markov shift is ergodic when the matrix Π is irreducible in the sense that for each pair (i, j) there exists an n such that $\pi_{ij}^{(n)} > 0$. The shift is mixing if there exists an n such that all of the elements of $\Pi^{(n)}$ are strictly positive (mixing implies irreducibility, and irreducibility plus aperiodicity implies mixing).

10.3 α-congruence

Equivalence implicit in isomorphism concerns the probabilistic structure of orbits, not their geometry, much of which can be distorted by the map that relates two isomorphic spaces. Therefore, geometrically different systems can be isomorphic. In order to overcome this difficulty the notion of α-congruence has been suggested. The following definition, based on Ornstein and Weiss (1991), pp. 22–3 and 63, is formulated in the more general terms of flows, but can be easily adapted to discrete-time systems.

definition 10.2 *Consider two flows ϕ_t and $\tilde{\phi}_t$, defined on the same metric space M and preserving the measures μ and $\tilde{\mu}$, respectively. We say that ϕ_t and $\tilde{\phi}_t$ are α-**congruent** if they are isomorphic via a map $\theta : M \to M$ that satisfies*

$$\mu\{x \mid d[x, \theta(x)] > \alpha\} < \alpha$$

where $x \in M$ and d is a fixed metric on M.

That is to say, the isomorphism θ moves points in M by less than α, except for a set of points in M of measure less than α. Thus, if α is so small that we do not appreciate distances in M smaller than α and we ignore events whose probability is smaller than α, then we consider α-congruent systems as actually indistinguishable. α-congruence is a form of equivalence stronger than isomorphism, as it requires that systems be close not only in a measure-theoretic sense but also in a geometric sense.

If we want to compare stochastic processes and deterministic systems, and the realisations of the former do not lie in the same metric space as the orbits of the latter, we need a definition of α-congruence covering the general case of two flows (or maps) acting on different abstract measure spaces (see Ornstein and Weiss 1991, pp. 22–5).

definition 10.3 *Let ϕ_t and $\tilde{\phi}_t$ be two flows acting on measure spaces M and $\tilde{M}$ and preserving the probability measures μ and $\tilde{\mu}$, respectively. Let $\mathcal{P}$ and $\tilde{\mathcal{P}}$ be two functions from M (or $\tilde{M}$) to the same metric space (X, d). We say that ϕ_t and $\tilde{\phi}_t$ are α-congruent if there exists an invertible measure-preserving map $\theta : M \to \tilde{M}$ such that*

(i) *$\theta \circ \phi_t = \tilde{\phi}_t \circ \theta$ almost everywhere;*
(ii) *for all measurable subsets $\tilde{A} \subset \tilde{X}$, $\mu\left[\theta^{-1}(\tilde{A})\right] = \tilde{\mu}(\tilde{A})$ (therefore, ϕ_t and $\tilde{\phi}_t$ are isomorphic); and*
(iii) *$d\left(\mathcal{P}(x), \tilde{\mathcal{P}}[\theta(x)]\right) < \alpha$ except for a set of μ-measure $< \alpha$.*

This definition can be easily adapted to maps. Suppose now that we have a discrete-time deterministic system defined by (X, G, ρ) and a stochastic process defined by (Ω, T, μ) where T is the shift map. A natural choice for the function $\tilde{\mathcal{P}}$ on the space of sequences Ω is the partition of Ω giving the state of the sequence at time zero, i.e., the current state of the process. The typical choice for $\mathcal{P}$ is a partition of the state space of the deterministic dynamical system that can be interpreted as an observation function such that each element of $\mathcal{P}$ includes points of the state space which cannot be distinguished at a certain level of precision of observation. If the functions $\tilde{\mathcal{P}}$ and $\mathcal{P}$ have the same range, realisations of the stochastic process and orbits of the deterministic system, observed with finite precision, can be compared. If the stochastic process and the deterministic system thus partitioned are α-congruent, in the sense of definition 10.3, we can say that the realisations of the former and the orbits of the latter are observationally indistinguishable at a level of precision depending on α.

The following, recently established results (Radunskaya 1992) shed some doubt on the possibility of distinguishing deterministic chaos and randomness observationally.

proposition 10.6 *Let ϕ_t be a flow on a manifold M that is isomorphic to the Bernoulli flow of infinite entropy. Then, for any $\alpha > 0$, there is a continuous time, finite state Markov process $\mathcal{M}_t$ taking value on M which is α-congruent to ϕ_t.*

This is a remarkable result but is not sufficient for our present purposes. In the study of deterministic chaotic systems, we usually consider the dynamics of the system on a compact attractor whose entropy is finite and bounded by the sum of the positive Lyapunov characteristic exponents. Therefore we are interested here in flows with finite entropy. The following result is then more relevant to our purpose.

proposition 10.7 *Let ϕ_t be a B-flow of finite entropy on a manifold M. Let B_t^∞ be an infinite entropy B-flow on a probability space Ω. Then, for any $\alpha > 0$, there exists a continuous-time Markov process $\mathcal{M}_t$, on a finite number of states $\{s_i\} \in M \times \Omega$ such that $\mathcal{M}_t$ is α-congruent to $\tilde{\phi}_t = \phi_t \times B_t^\infty$.*

These rather abstract results can be given a striking common-sense interpretation. Let us consider an observer looking at orbits generated by a deterministic Bernoulli system with infinite entropy and let us suppose that observation takes place through a device (a viewer) that distorts by less than α, with probability greater than $1 - \alpha$, where α is a positive number that we can choose as small as we wish. Proposition 10.6 (infinite entropy case) tells us that the orbits as seen through the viewer are arbitrarily close to a continuous-time, finite state Markov process. In the finite entropy case, in order to compare the orbits of the deterministic system with the sample paths of the (infinite entropy) Markov process, we need to introduce some additional entropy by 'sprinkling' the deterministic system with a bit of noninterfering noise. We can again use the parable of orbits observed through a slightly distorting viewer, but now the errors are random. Proposition 10.7 tells us that in this case, too, the observed orbits are most of the time arbitrarily close to the sample paths of the Markov process.

These sharp results should produce some scepticism on the possibility of rigorously testing whether a given series has been generated by a deterministic or a stochastic mechanism, for example, by estimating the value of correlation dimension or the dominant Lyapunov characteristic exponent. In view of propositions 10.6 and 10.7, if those tests are applied to a Markov process and to a deterministic Bernoulli system which are α-congruent, they should give the same results for sufficiently small values of α.

Our main concern is with the general consequences of the results above, and we leave details aside. However, a simple example of α-congruence is in order and we consider one between a discrete-time deterministic system and a discrete-time stochastic (Markov) process, using once again the tent map G_Λ, with the Lebesgue measure on the interval, which we know to be Bernoulli. For the tent map along with a partition of the state space, we can define a Markov process on k states (the number of states depending on α), such that its sample paths are α-indistinguishable from the orbits of the map.[4] For example, put $\alpha = 1/2$ and choose the partition $\mathcal{P} : [0,1] \to \{L, R\}$ where $L = \{x \in [0, 1/2]\}$, $R = \{x \in (1/2, 1]\}$. Then a Markov

---

[4]This example has been suggested by A. Radunskaya in a private correspondence with one of the authors.

process on these two states L and R, with transition matrix

$$
\begin{array}{cc}
 & \begin{array}{cc} L & R \end{array} \\
\begin{array}{c} L \\ R \end{array} &
\left(\begin{array}{cc} 1/2 & 1/2 \\ 1/2 & 1/2 \end{array} \right)
\end{array}
$$

will generate sample paths α-indistinguishable from those of the deterministic map G_Λ.[5]

If we reduce α and put, say, $\alpha = 2^{-k}, k \geq 2$, and choose a partition $\mathcal{P}_k$ such that the unit interval is divided into 2^k subintervals of equal length 2^{-k} defined by

$$
\left\{ x \left| \frac{j-1}{2^k} < x < \frac{j}{2^k} \right. \right\} \quad 1 \leq j \leq 2^k
$$

we can construct a 2^k state Markov process with transition matrix (omitting the state labels)

$$
\begin{pmatrix}
1/2 & 1/2 & 0 & 0 & \cdots & \cdots & 0 & 0 \\
0 & 0 & 1/2 & 1/2 & \cdots & \cdots & 0 & 0 \\
\cdots & \cdots & \cdots & \cdots & \cdots & \cdots & \cdots & \cdots \\
0 & 0 & 0 & 0 & \cdots & \cdots & 1/2 & 1/2 \\
\cdots & \cdots & \cdots & \cdots & \cdots & \cdots & \cdots & \cdots \\
0 & 0 & 1/2 & 1/2 & \cdots & \cdots & 0 & 0 \\
1/2 & 1/2 & 0 & 0 & \cdots & \cdots & 0 & 0
\end{pmatrix}.
$$

Again, the sequences generated by the Markov process can be matched (within α) to the orbits generated by the tent map on the k-partitioned space.

Our considerations so far can be given a more or less optimistic interpretation according to one's point of view and temperament. The results discussed above indicate that we cannot hope to provide a generally valid test for distinguishing deterministic chaos and true randomness. This would certainly be impossible for Bernoulli systems and, to the extent that the conjecture that 'most observable chaos is Bernoulli'[6] is correct, it would be generally impossible. Consequently, at least for a certain class of concrete dynamical systems, the possibility exists of representing them either

---

[5] The partition $\tilde{\mathcal{P}}$ on the space of sequences Ω is again chosen so that for each sequence it returns its current value in the space $\{L, R\}$, endowed with the discrete metric

$$
d(x, y) = \begin{cases} 0, & \text{if } x = y; \\ 1, & \text{if } x \neq y. \end{cases}
$$

[6] See Ornstein and Weiss (1991), p. 22.

as deterministic systems (plus perhaps some random, noninterfering disturbances), or as stochastic processes. The choice between the two is a matter
of expedience rather than theory. In principle, a deterministic representation is superior for the purpose of explanation, but this is only true if the
resulting model is sufficiently simple and we can provide a physical interpretation of the state variables and the functional relationships among them.

Further reading

The literature on nonlinear dynamical systems is vast and rapidly expanding. We shall not try to provide an exhaustive bibliography, but simply select a number of reference items, mostly books, that extend or complement the material presented in this book. The reader should keep in mind, however, that many important results in this field and most detailed analyses are available only in articles published in specialised journals. The selection that follows is divided into five broad and partially overlapping sections.

DIFFERENTIAL AND DIFFERENCE EQUATIONS

There exists a huge number of books on ordinary differential equations. Two classical texts are Arnold (1973) and Hirsch and Smale (1974). A more recent one, at a comparable level of difficulty, is Hale and Koçak (1991). Perko (1991), Braun (1993) and Blanchard *et al.* (1998) provide a coverage of the same area at an intermediate level of difficulty. Comprehensive texts on difference equations are fewer: we suggest Kelley and Peterson (1991).

STABILITY

Stability is one of the central questions in the classical theory of dynamical systems. Early fundamental results were published at the end of the nineteenth century, and important contributions in this field were subsequently provided by mathematicians, physicists, engineers, economists and other scientists, some of which are included in standard texts on differential or difference equations. In recent years, although researchers' interest in stability *per se* seems to have declined, the discussion of basic concepts of modern dynamical system theory, such as attracting sets and attractors, has necessarily involved many ideas and results of stability theory.

A serious investigation of stability cannot omit the study of the classical books by Hahn (1963), Hahn (1967), Bhatia and Szegö (1970) and the two survey articles by Kalman and Bertram (1960) are fundamental texts on Lyapunov's direct method.

DYNAMICAL SYSTEMS, BIFURCATIONS AND CHAOS

The rate of growth of the literature in this area since the 1970s has been phenomenal and selecting a small number of items is a particularly hard chore.

Guckenheimer and Holmes (1983) provides a somewhat dated treatment of nonlinear dynamics and chaos, but it has been a classical reference for twenty years now and still is inspiring reading.

The two recent books by Alligood *et al.* (1997) and Robinson (1999) contain exhaustive, advanced mathematical treatments of nonlinear dynamics, with rigorous proofs of the most important theorems and plenty of exercises. A very good exercise for the reader is to try and make sense of the different ways in which identically named concepts (e.g., attractors) are defined by these authors. It is unfortunate that Robinson's otherwise excellent book has been edited in a less than perfect manner.

Katok and Hasselblatt (1995) is a dense and difficult book covering most of the same ground as the ones mentioned above (and much more), with emphasis on the ergodic properties of dynamical systems. It is an invaluable reference text with many illuminating examples and exercises.

At a lower level of mathematical difficulty, Wiggins (1990) provides a clear and fairly complete coverage of the subject-matter, with emphasis on the geometrical properties of chaos and many exercises and examples. Wiggins' book also contains a particularly detailed discussion of homoclinic and heteroclinic bifurcations.

Devaney (1986) is a standard reference in this field. Its scope is much narrower than that of Alligood's or Robinson's books, because it deals mostly with one- and two-dimensional nonlinear maps. However, as an introductory but rigorous treatment of chaotic dynamical systems, this book beats most of the competition. The same author has published an even simpler introductory text, Devaney (1992), with similar virtues (and limitations). Another quite useful introductory text on chaos in low-dimensional discrete-time systems is Elaydi (2000), which contains some recent interesting results and provides clear demonstrations and many exercises with answers. Arrowsmith and Place (1992) and Glendinning (1994) nicely complement the books by Devaney and Elaydi. The two volumes of Peitgen *et al.* (1992) provide an excellent discussion of the many details involved in chaotic dynamics with emphasis on numerical and graphical methods.

The most complete and thorough treatments of two-dimensional dynamical systems in discrete time are probably Mira (1987) and Mira *et al.* (1996). Abraham *et al.* (1997) provides a very insightful pictorial presentation of two-dimensional dynamics.

Ott (1993) and Hilborn (1994) are books written by scientists and addressed to scientists and engineers. The coverage is broad, including items like quantum chaos, multifractals and chaotic scattering, and the style is pedagogic. Ott's book is shorter, sharper and its editing and graphics are far superior to those of Hilborn's.

ERGODIC THEORY

Billingsley (1965) and Cohen (1980) are two excellent reference books on measure theory. Doob (1994) is a short and brilliant, though somewhat idiosyncratic' book, especially useful for readers with a statistical or probabilistic background.

There are many good introductions to ergodic theory. Walters (1982) and Cornfeld *et al.* (1982) cover all the most important topics. Keller (1998) is a small, agile volume with a clear, rigorous presentation of the basic concepts and ideas. The survey article by Eckmann and Ruelle (1985) is also an extremely rewarding piece of reading.

The survey in Ornstein and Weiss (1991) regarding statistical properties of dynamical systems is mandatory reading for students who want to investigate further the notions of Bernoulli systems or α-congruence, although the authors' presentation is not always as clear and well organised as one would wish.

DYNAMICAL SYSTEM THEORY AND ECONOMICS

An early, very successful book on methods of economic dynamics by Gandolfo (which has been revised as Gandolfo (1996)) contains an introductory mathematical discussion of differential and difference equations and numerous applications to economics. Stokey and Lucas (1989) is a standard advanced textbook on iterative methods in economic dynamics, including both deterministic and stochastic models.

Among the books on nonlinear dynamical systems and chaos written by economists and including economic applications, we would like to mention: Brock and Malliaris (1989), Chiarella (1990), Goodwin (1990), Medio (1992), Lorenz (1993), Day (1994), Shone (1997), Puu (1997), Puu (2000). Survey articles or collections of readings discussing economic applications of chaos theory include: Baumol and Benhabib (1989), Boldrin and Woodford (1990), Scheinkman (1990), Benhabib (1992), Jarsulic (1993), Medio (1998, 1999).

Economic examples of applications of 'one-hump' maps include, among others: macroeconomic models, for example, Stutzer (1980), Day (1982); models of rational consumption, for example, Benhabib and Day (1981); models of overlapping generations, for example, Benhabib and Day (1982), Grandmont (1985); models of optimal growth, for example, Deneckere and Pelikan (1986). Overviews of the matter with further instances of 'one-hump' functions derived from economic problems can be found in Baumol and Benhabib (1989), Boldrin and Woodford (1990) and Scheinkman (1990). For a continuous-time generalisation of one-dimensional maps, see Invernizzi and Medio (1991).

We can mention three classes of economic models where quasiperiodicity occurs: models describing optimal growth in a discrete-time setting, for example, Venditti (1996); models of overlapping generations with production, for example, Reichlin (1986), Medio (1992); models of Keynesian (or perhaps Hicksian) derivation, describing the dynamics of a macroeconomic system characterised by nonlinear multiplier-accelerator mechanisms, for example, Hommes (1991). For a recent application of the concept of hysteresis to a problem of labour economics see Brunello and Medio (1996).

Bibliography

ABARBANEL, H.D.I., BROWN, R., SIDOROWICH, J.J. AND TSIMRING, L. SH. 1993. The analysis of observed chaotic data in physical systems. *Reviews of Modern Physics* **65**, 1331–92.

ABRAHAM, R. H., GARDINI, L. AND MIRA, C. 1997. *Chaos in Discrete Dynamical Systems: A Visual Introduction in Two Dimensions.* Santa Clara, CA: Springer-Vergag.

ABRAHAM, R. H. AND MARSDEN, J. E. 1978. *Foundations of Mechanics.* Reading, MA: Benjamin/Cummings.

ABRAHAM, R. H. AND SHAW, C. D. 1982, 1983, 1984, 1988. *Dynamics - The Geometry of Behaviour. Part 1, Part 2, Part 3, Part 4.* Santa Cruz, CA: Aerial.

ALLIGOOD, K. T., SAUER, T. D. AND YORKE, J. A. 1997. *Chaos - An Introduction to Dynamical Systems.* New York: Springer-Verlag.

ARNOLD, V. I. 1965. Small denominators I: mapping of the circumference into itself. *AMS Translation Series 2* **46**, 213.

ARNOLD, V. I. 1973. *Ordinary Differential Equations.* Cambridge, MA: MIT Press.

ARNOLD, V. I. 1980. *Chapitres Supplémentaires de la Théorie des Équations Différentielles Ordinaires.* Moscow: MIR.

ARROW, K. J., BLOCK, H. D. AND HURWICZ, L. 1959. On the stability of the competitive equilibrium, II. *Econometrica* **27**, 82–109.

ARROW, K. J. AND HURWICZ, L. 1958. On the stability of the competitive equilibrium, I. *Econometrica* **26**, 522–52.

ARROW, K. J. AND HURWICZ, L. 1960. Competitive stability under weak gross substitutability. The Euclidean distance approach. *International Economic Review* **1**, 38–49.

ARROW, K. J. AND KURZ, M. 1970. *Public Investment, the Rate of Return, and Optimal Fiscal Policy.* Baltimore: Johns Hopkins Press.

ARROWSMITH, D. K. AND PLACE, C. M. 1992. *Dynamical Systems: Differential Equations, Maps and Chaotic Behaviour.* London: Chapman & Hall.

BALA, V. AND MAJUMDAR, M. 1992. Chaotic tâtonnement. *Economic Theory* **2**, 437–45.

BANKS, J., BROOKS, J., CAIRNS, G., DAVIS, G. AND STACEY, P. 1992. On Devaney's definition of chaos. *American Mathematical Monthly* **99**, 332–4.

BAUMOL, W. J. AND BENHABIB, J. 1989. Chaos: significance, mechanism, and economic applications. *Journal of Economic Perspectives* **3**, 77–107.

BECK, D. AND SCHLÖGL, F. 1993. *Thermodynamics of Chaotic Systems.* Cambridge: Cambridge University Press.

BENDICKS, M. AND CARLESON, L.1991. The dynamics of the Hénon map. *Annals of Mathematics* **133**, 73–169.

BENHABIB, J. (ed.) 1992. *Cycles and Chaos in Economic Equilibrium.* Princeton: Princeton University Press.

BENHABIB, J. AND DAY, R. H. 1981. Rational choice and erratic behaviour. *Review of Economic Studies* **48**, 459–71.

BENHABIB, J. AND DAY, R. H. 1982. A characterization of erratic dynamics in the overlapping generations models. *Journal of Economic Dynamics and Control* **4**, 37–55.

BENVENISTE, L. M. AND SCHEINKMAN, J. A. 1979. On the differentiability of the value function in dynamic models of economics. *Econometrica* **47**, 727–32.

BERGÉ, P., POMEAU, Y. AND VIDAL, C. 1984. *Order within Chaos: Towards a Deterministic Approach to Turbulence.* New York: Wiley.

BERMAN, A. AND PLEMMONS, R. J. 1979. *Nonnegative Matrices in the Mathematical Sciences.* New York: Academic Press.

BHATIA, N. P. AND SZEGÖ, G. P. 1970. *Stability Theory of Dynamical Systems.* Berlin: Springer-Verlag.

BILLINGSLEY, P. 1965. *Ergodic Theory and Information.* New York: John Wiley.

BILLINGSLEY, P. 1979. *Probability and Measure.* New York: John Wiley.

BLANCHARD, P., DEVANEY, R. L. AND HALL, G. R. 1998. *Differential Equations.* London: Brooks/Cole.

BISCHI, G., MIRA, C. AND GARDINI, L. 2000. Unbounded sets of attraction. *International Journal of Bifurcation and Chaos* **10**, 1437–69.

BOLDRIN, M. AND WOODFORD, M. 1990. Equilibrium models displaying endogenous fluctuations and chaos: a survey. *Journal of Monetary Economics* **25**, 189–223.

BOYARSKY, A. AND GÓRA, P. 1997. *Laws of Chaos. Invariant Measures and Dynamical Systems in One Dimension.* Boston: Birkhäuser.

BRAUN, M. 1993. *Differential Equations and Their Applications: An Introduction to Applied Mathematics.* New York: Springer-Verlag.

BRIGGS, J. 1994. *Fractals: The Patterns of Chaos.* London: Thames & Hudson.

BROCK, W. A., DECHERT, W. D., SCHEINKMANN J. AND LEBARON, B. 1996. A test for independence based upon the correlation dimension. *Econometric Review* **15**, 197–235.

BROCK, W. A. AND MALLIARIS, A. G. 1989. *Differential Equations, Stability and Chaos in Dynamic Economics.* Amsterdam: North-Holland.

BRUNELLO, G. AND MEDIO, A. 1996. A job competition model of workplace training and education. *Working Paper* 96.16, Department of Economics, University of Venice.

CASS, D. AND SHELL, K. 1976a. The structure and stability of competitive dynamical systems. *Journal of Economic Theory* **12**, 31–70.

CASS, D. AND SHELL, K. 1976b. Introduction to Hamiltonian dynamics in economics. *Journal of Economic Theory* **12**, 1–10.

CHIANG, A. C. 1992. *Elements of Dynamic Optimization.* New York: McGraw-Hill.

CHIARELLA, C. 1990. *The Elements of a Nonlinear Theory of Economic Dynamics*. Berlin: Springer-Verlag.

COHEN, D. L. 1980. *Measure Theory*. Boston: Birkhäuser.

COLLET, P., ECKMANN, J.P. AND LANFORD, O.E. 1980. Universal properties of maps on an interval. *Communications of Mathematical Physics* **76**, 211–54.

CORNFELD, I. P., FOMIN, S. V. AND SINAI, YA. G. 1982. *Ergodic Theory*. New York: Springer-Verlag.

CORNFELD, I. P. AND SINAI, YA. G. 2000. General ergodic theory of groups of measure preserving transformations, in Ya. G. Sinai (ed.), *Dynamical Systems, Ergodic Theory and Applications*. Berlin: Springer-Verlag.

CVITANOVIĆ, P. 1984. *Universality in Chaos: A Reprint Selection*. Bristol: Adam Hilger.

DAY, R. H. 1982. Irregular growth cycles. *American Economic Review* **72**, 406–14.

DAY, R. H. 1994. *Complex Economic Dynamics. Volume 1. An Introduction to Dynamical Systems and Market Mechanisms*. Cambridge, MA: MIT Press.

DECHERT, W. D. (ed.) 1996. *Chaos Theory in Economics: Methods, Models and Evidence*. Cheltenham: Edward Elgar.

DENECKERE, R. AND PELIKAN, S. 1986. Competitive Chaos. *Journal of Economic Theory* **40**, 13–25.

DEVANEY, R. 1986. *Introduction to Chaotic Dynamical Systems*. Menlo Park, CA: Benjamin Cummings.

DEVANEY, R. L. 1992. *A First Course in Chaotic Dynamic Systems: Theory and Experiment*. Reading, MA: MIT Press.

DOOB, J. L. 1994. *Measure Theory*. New York: Springer-Verlag.

DUGUNDJI, J. 1966. *Topology*. Boston: Allyn & Bacon.

ECKMANN, J. P. 1981. Roads to turbulence in dissipative dynamical systems. *Reviews of Modern Physics* **53**, 643–54.

ECKMANN, J. P. AND RUELLE, D. 1985. Ergodic theory of chaos and strange attractors. *Reviews of Modern Physics* **57**, 617–56.

EDGAR, G. A. 1990. *Measure, Topology, and Fractal Geometry*. New York: Springer-Verlag.

ELAYDI, S. N. 2000. *Discrete Chaos*. Boca Raton, FL: Chapman & Hall/CRC Press.

FEIGENBAUM, M. J. 1978. Quantitative universality for a class of nonlinear transformations. *Journal of Statistical Physics* **19**, 25–52.

FRANCESCHINI, V. AND TIBALDI, C. 1979. Sequences of infinite bifurcations and turbulence in a truncation of the Navier–Stokes equations. *Journal of Statistical Physics* **21**, 707.

GANDOLFO, G. 1996. *Economic Dynamics*, 3rd edn. New York: Springer-Verlag.

GLENDINNING, P. 1994. *Stability, Instability and Chaos: An Introduction to the Theory of Nonlinear Differential Equations*. Cambridge: Cambridge University Press.

GOODWIN, R. M. 1967. A growth cycle, in C. H. Feinstein (ed.), *Socialism, Capitalism and Economic Growth*. Cambridge: Cambridge University Press. Revised version in E. K. Hunt and J. G. Schwarz (eds.), 1969 *A Critique of Economic Theory*. Harmondsworth: Penguin.

GOODWIN, R. M. 1990. *Chaotic Economic Dynamics*. Oxford: Clarendon Press.

GRANDMONT, J. M. 1985. On endogenous competitive business cycles. *Econometrica* **53**, 995–1045.

GRASSBERGER, P. AND PROCACCIA, I. 1983. Measuring the strangeness of strange attractors. *Physica D* **9**, 189–208.

GUCKENHEIMER, J. AND HOLMES, P. 1983. *Nonlinear Oscillations, Dynamical Systems, and Bifurcations of Vector Fields*. New York: Springer-Verlag.

HAHN, W. 1963. *Theory and Applications of Liapunov's Direct Method*. Englewood Cliffs NJ: Prentice-Hall.

HAHN, W. 1967. *Stability of Motion*. Berlin: Springer-Verlag.

HALE, J. AND KOÇAK, H. 1991. *Dynamics and Bifurcations*. New York: Springer-Verlag.

HARTMAN, P. 1964. *Ordinary Differential Equations*. New York: John Wiley.

HASSARD, B. D., KAZARINOFF, N. D. AND WAN, Y.-H. 1980. *Theory and Applications of the Hopf Bifurcation*. Cambridge: Cambridge University Press.

HILBORN, R. C. 1994. *Chaos and Nonlinear Dynamics: An Introduction for Scientists and Engineers*. Oxford: Oxford University Press.

HIRSCH, M. W. AND SMALE, S. 1974. *Differential Equations, Dynamical Systems and Linear Algebra*. New York: Academic Press.

HOMMES, C. H. 1991. *Chaotic Dynamics in Economic Models. Some Simple Case-Studies*. Groningen: Walters-Noordhoff.

INVERNIZZI, S. AND MEDIO, A. 1991. On lags and chaos in economic dynamic models. *Journal of Mathematical Economics* **20**, 521–50, reprinted in Dechert (1996).

IOOSS, G. 1979. *Bifurcation of Maps and Applications*. Amsterdam: North-Holland.

JACKSON, E.A. 1990. *Perspectives of Nonlinear Dynamics*, 2 vols. Cambridge: Cambridge University Press.

JAKOBSON, M. V. 2000. Ergodic theory of one-dimensional mappings, in Ya. G. Sinai (ed.), *Dynamical Systems, Ergodic Theory and Applications*. Berlin: Springer-Verlag.

JARSULIC, M. (ed.) 1993. *Nonlinear Dynamics in Economic Theory*. Cambridge: Edward Elgar.

KALMAN, R. E. AND BERTRAM, J. E. 1960. Control system analysis and design via the 'second method' of Lyapunov I, II. *Transactions ASME Series D. Journal of Basic Engineering* **82**, 371–93, 394–400.

KAPLAN, J. L. AND YORKE, J. A. 1979. *Chaotic Behaviour of Multidimensional Difference Equations*. Berlin: Springer-Verlag.

KATOK, A. AND HASSELBLATT, B. 1995. *Introduction to the Modern Theory of Dynamical Systems*. Cambridge: Cambridge University Press.

KELLER, G. 1998. *Equilibrium States in Ergodic Theory*. Cambridge: Cambridge University Press.

KELLEY, W. G. AND PETERSON, A. C. 1991. *Difference Equations: An Introduction with Applications*. Boston: Academic Press.

KRASOVSKII, N. N. 1963. *Stability of Motions*. Palo Alto, CA: Stanford University Press.

LANFORD, O. E. 1982. A computer assisted proof of the Feigenbaum conjectures. *Bulletin of the American Mathematical Society* **6**, 427–34.

LASOTA, A. AND MACKEY, M. C. 1985. *Probabilistic Properties of Deterministic Systems*. Cambridge: Cambridge University Press.

LI, T. Y. AND YORKE, J. A. 1975. Period three implies chaos. *American Mathematical Monthly* **82**, 985–92.

LORENZ, E. 1963. Deterministic nonperiodic flow. *Journal of Atmospheric Sciences* **20**, 130–41.

LORENZ, H. W. 1993. *Nonlinear Dynamical Economics and Chaotic Motion*, 2nd edn. Berlin: Springer-Verlag.

MACKAY, R. S. 1992. Nonlinear dynamics in economics: a review of some key features of nonlinear dynamical systems. Florence: European University Institute, manuscript.

MAGILL, M. J. P. 1977. Some new results on the local stability of the process of capital accumulation. *Journal of Economic Theory* **15**, 174–210.

MANDELBROT, B. B. 1982. *The Fractal Geometry of Nature*. San Francisco, CA: Freeman.

MAÑE, R. 1987. *Ergodic Theory and Differentiable Dynamics*. Berlin: Springer-Verlag.

MEDIO, A. 1992. *Chaotic Dynamics. Theory and Applications to Economics*. Cambridge: Cambridge University Press.

MEDIO, A. 1998. Nonlinear dynamics and chaos. Part I: a geometrical approach. *Macroeconomic Dynamics* **2**, 505–32.

MEDIO, A. 1999. Nonlinear dynamics and chaos. Part II: ergodic approach. *Macroeconomic Dynamics* **3**, 84–114.

MEDIO, A. 2001. The problem of backward dynamics in economic models. *Working paper* 2001.01, Department of Economics, University of Venice.

METZLER, L. 1945. Stability of multiple markets: the Hicks conditions. *Econometrica* **13**, 277–92.

MILNOR, J. 1985. On the concept of attractor. *Communications in Mathematical Physics* **99**, 177–95.

MIRA, C. 1987. *Chaotic Dynamics*. Singapore: World Scientific.

MIRA, C., GARDINI, L., BARUGOLA, A. AND CATHALA, J. C. 1996. *Chaotic Dynamics in Two-Dimensional Non-Invertible Maps*. Singapore: World Scientific.

MORI, H. AND KURAMOTO, Y. 1998. *Dissipative Structures and Chaos*. Berlin: Springer-Verlag.

NEGISHI, T. 1962. The stability of a competitive economy: a survey article. *Econometrica* **30**, 635–69.

NEIMARK, YU. I. 1959. On some cases of periodic motion depending on parameters. *Dokl. Akad. Nauk SSSR* **129**, 736–9.

NEIMARK, YU. I. AND LANDA, P. S. 1992. *Stochastic and Chaotic Oscillations*. Dordrecht: Kluwer Academic Publishers.

NEWHOUSE, S., RUELLE, D. AND TAKENS, F. 1978. Occurrence of strange axiom A attractors near quasi periodic flows on T^m, $m \geq 3$. *Communications in Mathematical Physics* **64**, 35–40.

OONO, Y. AND OSIKAWA, M. 1980. Chaos in nonlinear difference equations, I. *Progress of Theoretical Physics* **64**, 54–67.

ORNSTEIN, D. S. 1974. *Ergodic Theory, Randomness, and Dynamical Systems*, Yale Mathematical Monographs 5. New Haven and London: Yale University Press.

ORNSTEIN, D. S. AND WEISS, B. 1991. Statistical properties of chaotic systems. *Bulletin (New Series) of the American Mathematical Society* **24**, 11-115.

OSLEDEC, V. I. 1968. A multiplicative ergodic theorem. Liapunov characteristic numbers for dynamical systems. *Transactions of the Moscow Mathematical Society* **19**, 197–221.

OTT, E. 1993. *Chaos in Dynamical Systems*. Cambridge: Cambridge University Press.

PALIS, J. AND TAKENS, F. 1993. *Hyperbolicity and Sensitive Chaotic Dynamics at Homoclinic Bifurcations*. Cambridge: Cambridge University Press.

PEITGEN, H.-O., JÜRGENS, H. AND SAUPE, D. 1992. *Fractals for the Classroom*. Part 1, Part 2. New York: Springer-Verlag.

PERKO, L. 1991. *Differential Equations and Dynamical Systems*. New York: Springer-Verlag.

PONTRYAGIN, L. S., BOLTYANSKII, V. G., GAMKRELIDZE, R. V. AND MISHCHENKO, E. F. 1962. *The Mathematical Theory of Optimal Processes*. New York: Wiley Interscience.

PUU, T. 1997. *Nonlinear Economic Dynamics*. Berlin: Springer-Verlag.

PUU, T. 2000. *Attractors, Bifurcations and Chaos, Nonlinear Phenomena in Economics*. Berlin: Springer-Verlag.

RADUNSKAYA, A. 1992. *Alpha-Congruence of Bernoulli Flows and Markov Processes: Distinguishing Random and Deterministic Chaos*, PhD thesis. Palo Alto: Stanford University, unpublished.

REICHLIN, P. 1986. Equilibrium cycles in an overlapping generations economy with production. *Journal of Economic Theory* **40**, 89–102.

ROBINSON, C. 1999. *Dynamical Systems: Stability, Symbolic Dynamics and Chaos*. Boca Raton, FL: CRC Press.

RÖSSLER, O. E. 1976. Chemical turbulence: chaos in a small reaction-diffusion system. *Z. Naturforsch* **31A**, 1168–72.

RUELLE, D. 1989. *Chaotic Evolution and Strange Attractors*. Cambridge: Cambridge University Press.

RUELLE, D. AND TAKENS, F. 1971. On the nature of turbulence. *Communications of Mathematical Physics* **20**, 167–92.

SACKER, R. S. 1965. On invariant surfaces and bifurcations of periodic solutions of ordinary differential equations. *Communications of Pure and Applied Mathematics* **18**, 717–32.

SCHEINKMAN, J. A. 1990. Nonlinearities in economic dynamics. *Economic Journal* **100**, 33–49.

SCHUSTER, H. G. 1989. *Deterministic Chaos: An Introduction*, 2nd edn. Weinheim: VCH.

SHAW, R. 1981. Strange attractors, chaotic behaviour, and information flow. *Z. Naturforsch* **36A**, 80–112.

SHILNIKOV, L. 1997. Homoclinic chaos, in E. Infeld, R. Żelazny and A. Gałkowski (eds.), *Nonlinear Dynamics, Chaotic and Complex Systems*. Cambridge: Cambridge University Press.

SHONE, R. 1997. *Economic Dynamics. Phase Diagrams and Their Economic Application*. Cambridge: Cambridge University Press.

SMALE, S. 1967. Differentiable dynamical systems. *Bulletin of American Mathematical Society* **73**, 747–817.

SPARROW, C. 1982. *The Lorenz Equations: Bifurcations, Chaos, and Strange Attractors*. New York: Springer-Verlag.

STOKEY, N. L. AND LUCAS, R. E. 1989. *Recursive Methods in Economic Dynamics*. Cambridge, MA: Harvard University Press.

STUTZER, M. J. 1980. Chaotic dynamics and bifurcation in a macro-model. *Journal of Economic Dynamics and Control* **2**, 353–76.

SUTHERLAND, W. A. 1999. *Introduction to Metric and Topological Spaces.* Oxford: Oxford University Press.

THOMPSON, J. M. T. AND STEWART, H. B. 1986. *Nonlinear Dynamics and Chaos.* Chichester: Wiley.

TURNOVSKY, S. J. 1996. *Methods of Macroeconomic Dynamics.* Cambridge, MA: MIT Press.

VELLEKOOP, M. AND BERGLUND, R. 1994. On intervals: transitivity → chaos. *American Mathematical Monthly* **101**, 353–5.

VENDITTI, A. 1996. Hopf bifurcation and quasi-periodic dynamics in discrete multisector optimal growth models. *Research in Economics/Ricerche Economiche* **50**, 267–91.

WALRAS, L. 1874. *Eléments d' Economie Politiques Pure.* Lausanne: L. Corbaz. Translated by William Jaffe (1954) as *Elements of Pure Economics.* Homewood, Ill.: Richard D. Irwin.

WALTERS, P. 1982. *An Introduction of Ergodic Theory.* Berlin: Springer-Verlag.

WHITLEY, D. 1983. Discrete dynamical systems in dimensions one and two. *Bulletin of the London Mathematical Society* **15**, 177–217.

WIGGINS, S. 1988. *Global Bifurcations and Chaos: Analytical Methods.* New York: Springer-Verlag.

WIGGINS, S. 1990. *Introduction to Applied Nonlinear Dynamical Systems and Chaos.* New York: Springer-Verlag.

WIMMER, H. W. 1984. The algebraic Riccati equation: conditions for the existence and uniqueness of solutions. *Linear Algebra and its Applications* **58**, 441–52.

WOODFORD, M. 1986. Stationary sunspot equilibria in a finance constrained economy. *Journal of Economic Theory* **40**, 128–37.

Subject index

9 780521 551861